"This book gives billions and billions of reasons why pseudoscience is just plain wrong!"
— Kristine Larsen, Geological Sciences, Central Connecticut State University

"A brilliant book, revolutionary and world-changing. Oh, and *Doughnuts of the Gods?* That one was interesting too."
—Arnold Einstein, NotaNobel Laureate

"Deliciously explodes the myths of pseudoscience."
—Romuald Byczkiewicz, Historian, Central Connecticut State University

"Where did we park the car again?"
—Mrs. Agnes Thompson

"This book will hopefully make you reconsider everything authors such as Erich von Däniken, Immanuel Velikovsky, Graham Hancock, and Zecharia Sitchin have ever said."
—Most Any Reputable Scholar, Earth

"People should talk about this book as much as they talk about *Chariots of the Gods.*"
—Some Guy

"Vedeler and Silly have made a classical error in their hypothesis. Doughnuts? Surely not. The alien ships were in the shape of bagels. They came here in search of smoked salmon when the Lox Mines of Proxima Centauri closed. If that makes any sense to you, you definitely need to read this book. Immediately. That, or seek medical attention."
—Ken Feder, Archaeologist, Central Connecticut State University

"Mmmmmmm… Doughnuts…."
—Homer Simpson

"In this age of doubt about science, Dr. Vedeler skewers pseudo-science and its deluded minions."
— Steve McGrath, Historian, Central Connecticut State University

Also by Torger Vedeler

Intersect: A Love Story
Layers
Valley of Bones and Other Stories

DOUGHNUTS OF THE GODS

Torger Vedeler
and V.R.Y. Silly

Altakkme Books

DOUGHNUTS OF THE GODS

ISBN: 978-1-936783-15-1 (paperback)
ISBN: 978-1-936783-16-8 (e-book)

Altakkme Books
altakkme@gmail.com

For Marty, who taught me so much,

and

my students, who always inspire

CONTENTS

Introduction

In the spring of 1981, I had the opportunity to take a class at the University of New Mexico entitled "Gods, Heroes and Men," taught by Dr. Frank Wozniak, an archaeologist and historian and one of the finest teachers I've ever had. The class focused on unorthodox methods of understanding the past, particularly the ancient past, and included legendary material such as the Homeric epics, King Arthur, and Robin Hood, as well as views from what is commonly called the "fringe," including Immanuel Velikovsky, Atlantis, and the ancient astronaut beliefs of Erich von Däniken. This was a time when Velikovsky's theories of catastrophism were still very common on college campuses, and von Däniken's ideas were also near the peak of their popularity. As a boy in the late 1970s, I can still remember the excitement among my friends about the notion that our distant past had been populated by aliens, and the adventures of von Däniken as he pursued these things naturally fired all of our imaginations.

I also remember being skeptical. Something seemed very wrong with the ancient alien idea, though at the time I was unable to put my finger on quite what it was. I recall, however, wondering why not knowing how some ancient monuments were built could be seen as proof that they had been built by aliens. How did a negative assertion prove a positive claim, after all? I, like my friends, did not question von Däniken's data, though after taking Dr. Wozniak's course I realized that I should have. To put it mildly, von Däniken was a careless researcher, and some of his claims, such as his visit to a library of golden tablets in Ecuador, were apparently outright fabrications.[1]

1

This was my introduction to the phenomenon of "pseudoscience." As the name indicates, pseudoscience looks like science but is not. It is a mimic of science, and often involves a certain degree of intellectual dishonesty, sometimes but not always intentional. The term, among scholars of all sorts, is decidedly pejorative, and it is not something you call a colleague unless you want to start a fight. You may accuse them of practicing "bad science" or "bad scholarship," but this still implies that they are among the community of scientists, or of scholars.

Pseudoscientists, on the other hand, are the fringe, the "crackpots," and conflicts between them and scholars are often fierce and show no sign of slowing down, for even as individual pseudoscientific proposals fail, new ones always rise to take their place, making the whole enterprise rather like a game of "Whack-a-Mole." A part of the problem is that scholars have been unable to define precisely where science ends and pseudoscience begins, or even what pseudoscience exactly is. In his day, Heinrich Schliemann was considered a crackpot, a pseudoscientist, and yet his excavations at the mound of Hisarlik, with the invaluable help of Frank Calvert and Wilhelm Dörpfelt, revolutionized our understanding of the Homeric epics by demonstrating that Troy was, in fact, an actual place. To the fields of Greek archaeology and history, their discoveries constituted nothing less than a scientific revolution. Still, we must not read too much into the occasional success of those who are considered to be pseudoscientists. For every Schliemann, after all, there are thousands of von Dänikens, most of whose ideas do not flare brightly and who labor in obscurity. Von Däniken's unusual success is less for the quality of his ideas than their timing, since in the 1970s, the Apollo program had landed men on the moon and a future in space was both novel and exciting. It was an important time in science fiction, and aliens were in vogue.

Yet the problem remains: Where *is* the line between science and pseudoscience? This is called the "demarcation problem," and philosophers of science have been wrestling with it for many years.[2] Michael Shermer, for example, defines pseudoscience as *"claims pre-*

sented so that they appear scientific even though they lack supporting evidence and plausibility."[3] The problem with this definition is that it relies upon a value judgment about just what constitutes "evidence" and just what constitutes "plausibility." Mario Bunge argues that the difference between science and pseudoscience involves how they deal with "changeability, compatibility with the bulk of the antecedent knowledge, intersection with at least one other science, and control by the scientific community."[4] But while they accurately describe many of the features of pseudoscience, these definitions do not consider its underlying causes. To cite the example I am examining in this book, an ancient astronaut believer can produce a staggering amount of evidence, including antecedent evidence, to support their claims, and one feature of ancient visitations to Earth by extraterrestrials is that they are, strictly speaking, not impossible, given that all you would need for it to happen is the existence of an extraterrestrial civilization (something that astronomers take quite seriously these days) and their ability and desire to visit Earth. This was pointed out by Carl Sagan in 1963, before the explosion of interest in the subject that followed the publication of von Däniken's *Chariots of the Gods?* in 1968.[5] And yet, as I hope to demonstrate with V.R.Y. Silly's ancient doughnut "theory" and my analysis of it and similar proposals, ancient alien belief, as it is currently manifested in our society, is clearly a pseudoscience.[6]

The crux of the issue is not the existence of evidence or probability but rather how they are interpreted, even as this runs the risk of returning us to square one, for if a "correct" interpretation of data is the sole distinguishing characteristic of science, then is not the view of the pseudoscientist merely one possible explanation among many? How do we determine what is "correct"? Legitimate science, of course, favors those interpretations that better explain the evidence, but then how do we distinguish pseudoscience from mere bad science? As a solution to these circular arguments, I propose that pseudoscience differs from science in two major ways, both of which must be present if the term is to be justified; either alone merely means bad science or no science at all. First, pseudosciences exhibit extreme paradigm rigidity, and second, they confuse the truth value of *mythos* with that of *logos*, a distinction I will elaborate on below.

Paradigm rigidity is a question of more than simply the quality of interpretation (pseudoscientists tend to badly mishandle the evidence they present, but so do bad scientists), but also in identifying science (or more broadly, any field of scholarship that involves the interpretation of data) not as a *thing* or set of characteristics but rather as a social and intellectual *process,* following the approach laid out by Thomas Kuhn in his book *The Structure of Scientific Revolutions.*[7] Science, Kuhn argues, is based ultimately on paradigms, which are sets of statements about the functioning of phenomena within a field that are agreed upon by researchers in that field. In "normal science," researchers test the paradigm, either through experimentation or observation, and if these tests produce anomalies which cannot be resolved by the paradigm, a crisis can follow that results in the development of a new paradigm and then a "scientific revolution," or "paradigm shift," as the new paradigm replaces the old. Central to Kuhn's view is that a successful scientific revolution and paradigm shift explains not only the anomalies but *all* the data in the field, and that it will largely replace the older, less effective paradigm, resulting in progress in the field of study. Newton's laws of physics work well, really well, for example (you can go to the moon with them), but Einstein's explanations work better.

So what pseudoscientists claim to be doing (though few put it in these terms) is introducing a new paradigm into a field and initiating a scientific revolution.[8] Therefore, we are told, we should treat them as scientists, sometimes on a par with luminaries such as Darwin or Einstein, who actually did initiate real paradigm shifts. The problem for the pseudoscientist is that by its nature, pseudoscience does not progress or add significantly to the body of scientific knowledge, but rather tends to simply repeat itself, including its errors.[9] Indeed, a closer look at how pseudoscience works tells us that the pseudoscientist is actually using data quite differently from the scientist, and this I propose is an important element of the line of demarcation: like science, pseudoscience is not a thing but a process, but its process is not the same as science.

Both science and pseudoscience start with a period of inductive data gathering, which is to say the collection of data without a paradigm. In science, Kuhn refers to this as the "pre-paradigm period" in which there are several schools of thought that do not fully integrate with each other.[10] Both science and pseudoscience then form hypotheses and theories to explain the data, and both form paradigms with which they then address new data deductively. But here they diverge. Deductive science is designed to *test* its paradigms; its goal is to establish as objective a view and explanation of the evidence as possible, and all scientists are aware of the possibility of scientific revolutions, that there may be a better way of explaining the phenomena they are observing and interpreting. The paradigm in pseudoscience, on the other hand, grows rigid, often because it is reached too quickly, with not enough time spent in the initial inductive phase, so the data gathered is more likely to be superficial and without broader context (it may be as little as a single observation: "That pyramid is big!"). Unlike scientific paradigms, pseudoscientific paradigms do not explain, or even try to explain, all the data in a field. Rather, contradictory evidence is either ignored or attempts are made to discredit it, usually without making any real effort to understand what it says. This gives the pseudoscientist the advantage of never having to face or deal with anomalies, and as I noted above, the appearance of anomalies is essential to scientific revolutions and the evolution of science. Because its paradigm is so rigid, pseudoscience cannot therefore advance past the paradigm stage. The gathering of deductive data (and pseudoscientists often gather large quantities of this) is selective, because the pseudoscientist does not wish to test their theory but rather to *prove* it. As Ronald Fritze has observed, pseudoscientific beliefs are defended in much the same way as points in a legal case, since the goal is to win the argument, not to establish an objective understanding of the empirical world.[11] In other words, the first thing that distinguishes pseudoscience from science is that science evolves, but pseudoscience does not. A look at the histories of pseudosciences bears this out. Most, like Velikovskyism, last only as long as their founders,[12] and those, like creationism, that live longer do so because of other historical circumstances (in this case the religious anxiety that also produces religious fundamentalism).[13] Each pseudoscience, by its nature, tends toward the development of an orthodoxy around its paradigm, and enforcing and protecting that orthodoxy becomes paramount to its followers.

The second element of demarcation between science and pseudoscience involves the question of factual truth versus meaning. It is perfectly possible to find forms of truth in non-factual things and events, but we must keep in mind that these sorts of truths cannot be evaluated by scientific means. Science is by its nature empirical; it is concerned with facts and an understanding of the world that we can perceive in some way with our senses (or our senses enhanced with technology). This is a form of truth often described by the word *logos*. It is the type of truth you need if you are going to understand *how* the natural world works, but it is entirely neutral in terms of *why*. Scientists, of course, are human, and so can insert meaning into the data they observe. One can find the images produced by the Hubble Space Telescope beautiful, even religious, but that does not mean that the *data* we are seeing has those values; the meaning comes from us. Truths of meaning, often called *mythos*, are outside the purview of science. These are the "moral of the story," as it were, and while their truth value is just as real as that of *logos* truths, it is not the same thing and is not testable scientifically; to try and do so can render it sterile. *Mythos* truths are the sorts of truths we find in myths, in religious enlightenment, in moral parables, and in things like aesthetics. Their purpose is to bring significance to our lives, to explain *why* the universe is the way it is.[14] Unlike science, however, pseudoscience spends a great deal of time in *mythos*, often confusing it with *logos*. Many pseudosciences take on an almost religious character, and some, like creationism, are in fact derived not from science at all, but from religion, although unlike pseudoscience or pseudoreligion, normal religion *is* capable of evolution; this is in fact how it survives. Pseudosciences usually promise meaning to their adherents, and it is ironic that even as they claim to be searching for factual knowledge, the maintenance of an air of mystery is often an essential part of their success.

The original idea of extraterrestrial doughnuts came to me during Dr. Wozniak's fascinating class back in 1981, as a parody of the pseudosciences we discussed (I still have my very first handwritten draft, *Doughnuts from Heaven*, by "Henry Assume A Lot").[15] Ancient doughnut "theory" was therefore initially conceived as entertainment, a way to get a chuckle. Some years later, at Central Connecticut State University, I had a student come by my office and ask, quietly, what I thought about ancient aliens. I remember the visit well; I realized right away that the single worst thing I could do would be to call the idea stupid, since this would imply that those who might believe it or even consider it were stupid (and as we will see they are not; besides, no teacher should *ever* call a student stupid), and that this is actually the quickest way there is to ensure that a pseudoscientific notion gains adherents. So instead I told him it was an interesting idea (which it is), and we spent about an hour considering it. I showed him the Drake Equation and we discussed the possibility of life on other planets, the possibility of alien visitations to Earth, and the lack of positive evidence for this in the archaeological and historical record, leading the two of us to consider together the difference between positive and negative evidence. I also told him about my grandfather, who had worked on the Air Force investigations of UFOs in the 1960s (the Condon Report), and my own conversations with him about extraterrestrial life.[16]

Dr. Wozniak and this student ultimately led to *Doughnuts of the Gods*. The idea of ancient doughnuts provided a nice way to entertain my classes, and in 2011 I presented it at a science fiction conference in Boston as part of the "tinfoil hat" panel on crazy science fiction, and this got a lot of laughs. But the theory has more value than just comedy, and as I prepared to teach my own version of Dr. Wozniak's class in 2012, I realized that it could be part of a critical thinking exercise. For this I owe a debt of gratitude to Carl Sagan, who famously remarked in 1976: "I also hope for the continuing popularity of books like *Chariots of the Gods?* in high school and college logic courses, as object lessons in sloppy thinking. I know of no recent books so riddled with logical and factual errors as the works of von Däniken."[17]

It is my hope that *Doughnuts of the Gods* might serve in a similar vein, though its primary purpose is not to debunk any particular pseudoscientific theory.[18] The theory of intergalactic doughnuts is of course ridiculous, which is precisely the point. As an exercise I suggest the following: each of V.R.Y. Silly's statements has been assigned a number, which corresponds to a number in the Commentary. Reading Silly's text, the student should write down each logical flaw they find, as well as errors of omission, distortion and fact. As well, take note of the tone of the text, since I have tried to borrow the writing style from a number of pseudoscientific works. Since tone is a method of transmitting meaning (and therefore *mythos*), this part of any pseudoscientific work should be evaluated just as closely as what the pseudoscientist is claiming. Remember that pseudoscientists manipulate you not just with what they say or don't say, but *how* they say it. Have a look at the exciting presentation style of the *Ancient Aliens* series on the History Channel if you don't believe me.

Once you have analyzed the Silly text, turn to the corresponding section in the commentary and see my own critique of the ancient doughnut argument. You may find things I missed, and vise-versa, which is fine; the purpose of this whole thing is pedagogical, after all. Wherever possible I have included references to pseudoscientific works where similar mistakes, omissions and distortions have occurred, and I have tried to tie my commentary into the view of pseudoscience as a process that I outlined above. What was frightening to me as I put this thing together was how virtually no idea I came up with for Mr. Silly was so outrageous that it had not been proposed as a serious explanation of something by an actual pseudoscientist. Clearly among the legions of pseudoscientists I am but a rank amateur, and I stand humbly in their shadows.

I also learned something else while writing this book: Once I started looking for doughnuts in ancient texts and artifacts, *I started seeing them everywhere.* This is an important point, since it shows that when you expect to find something, whether it is ancient doughnuts or ancient aliens or Atlantis, you *will* find it. This is the power of the rigid paradigm, and it should serve as a cautionary tale to mainstream scholars as well. So let me close with a statement and a warning: Ancient Doughnut Theory is a parody and a satire, intended as comedy and to help teach critical thinking skills. *It is not true.* If you ignore me here and decide to believe that intergalactic space doughnuts actually built ancient monuments and mated with our ancestors and are planning to come back someday, I reserve the right to be the first to call you wrong.

Honestly, I'd rather share a laugh with you instead, and I hope that this book will give us both a reason to chuckle.

[1] See von Däniken 1974; Ferris 1974, 58; Story 1976, 87-90.

[2] See Gordin 2012, 7-14, who discusses the efforts and difficulties of earlier arguments over pseudoscience, including the criterion of "falsifiability" proposed by Karl Popper.

[3] 2002, 33. Italics in original.

[4] 2009, 238-239.

[5] See Shklovskii and Sagan, *Intelligent Life in the Universe* (1966) and von Däniken, *Chariots of the Gods?* (1970).

[6] The word "theory" merits a brief discussion. In scientific circles it is "the best tested, most consistent, working explanation for something based on a huge collection of evidence and tests. The word theory has come to be synonymous with hypothesis, or idea, or concept, or 'wild arse guess' in common vernacular." (Kristine Larsen, personal communication, 2019; see also Larsen 2008). As a historian I lean toward a broader definition, since in my field we are often forced to produce theories from small quantities of evidence and we cannot perform experimental tests on that evidence the way the hard sciences do. Either way, though, the fact that we are having this discussion and looking at how "theories" are seen by the wider public is decidedly a good thing.

[7] 1962. While this is a dry read, it is also essential not only for understanding science, but much about how human beings develop and refine information.

[8] Rand and Rose Flem-Ath are an exception here. They argue that Charles Hapgood's theory of polar shifts does constitute a scientific revolution (1995, 2012, 34). So far the rest of the community of geologists has not followed them.

[9] Shermer 2002, 38-41; Bunge 2009, 239.

[10] 1962, 48.

[11] 2009, 218.

[12] See Gordin 2012, 203.

[13] See Armstrong 2000, 178.

[14] For an excellent discussion of this question, see Armstrong 2000, xv-xviii.

[15] I suppose I should note that the idea that ancient doughnuts built the pyramids is different from the idea that the Earth is shaped like a doughnut, which some people apparently hold. See Mufson, 2018.

[16] See also Roach 1999, 168-173.

[17] Quote from Story 1976, 11.

[18] For that sort of thing the reader is referred to Story 1976, James and Thorpe 1999, and Feder 2002, among others.

One: The Great Cosmic Mystery

1.1

Ever since the beginnings of mankind and before, the human race has looked up at the night sky with wonder and awe. What mysteries it holds, what vastness! Our world—which to us is the entirety of our experience and all we can conceive of as real—seems so large as we go about our mundane tasks of life, as we are born and grow and live and die, as each generation passes and leaves its accumulated wisdom to those who come after. And each generation considers itself to be the last, that it and it alone has unlocked the secrets of the universe, the origins and fates of all things.

But is this truly so?

1.2

The Earth—our home—is in truth but a speck in the greater cosmos, a mote of dust spinning around an insignificant star near the edge of a galaxy that, if you were to see the universe as it is, would itself pass into anonymity. For scientists have estimated that there are over 100 billion stars in our galaxy alone, and in the universe, more than 200 *billion* galaxies! Simple calculations can be performed by anyone to come up with a total of 20,000 *billion* stars in the universe (20,000,000,000,000,000!). Recent discoveries of extrasolar planets (hundreds already, with more being discovered every day)[1] tell us that planets like Earth are common in the galaxy, and with them life must be common too. So if we assume that every

---

[1] Interested readers can follow this amazing progress at
http://planetquest.jpl.nasa.gov/, and witness the intense professionalism of NASA employees at
http://www.youtube.com/watch?v=QFvNhsWMU0c.

star has eight planets, this would mean that there are 160,000 *billion* planets (160,000,000,000,000,000) in the universe! Indeed, astronomers have calculated that the number of intelligent alien species in our own galaxy alone must be in excess of 100,000, based upon sophisticated statistical modeling. And math, as anyone knows, does not lie—numbers are absolutes, are fixed in their certainty. And so if there are 100,000 other intelligent races so close by, it is undeniable by even the simplest mind that extraterrestrials must have visited the Earth. Can we doubt this? These numbers are real, having been arrived at by the calculations of our most brilliant minds, men who have studied the problem for decades.

1.3

Yet so many deny that this could be! We are told by the "experts": *it could not happen.* Science has *spoken,* has *decided.* It has concluded that it and it alone holds the answers to the mysteries of the cosmos, and of the past, and by scholars saying that something is so, it becomes so. Those who question them, who dare to ask if perhaps things are not as the scholars say, they are condemned as cranks, as quacks, as crackpots. The evidence is consulted and a theory is made, and that theory is *fact!* It may not be challenged or changed, for to do so is to upend the entire edifice of carefully selected "truths" upon which scientists have based their Ivory Tower reputations, their comfortable positions, their beliefs. Albert Einstein once famously said: "Great spirits have always found violent opposition from mediocrities. The latter cannot understand it when a man does not thoughtlessly submit to hereditary prejudices, but honestly and courageously uses his intelligence and fulfills the duty to express the results of his thought in clear form."[2]

Should we not heed this wisdom spoken by the greatest mind of our times? Einstein refused to bow down to the ignorant, to the closed minds who told him that the evidence did not support him, that his ideas were too unorthodox to be true. And so they perse-

---

[2] Albert Einstein, quoted in the *New York Times,* March 19, 1940.

cuted him; he was forced to flee his native land to a country less prejudiced, where he could make his discoveries and change the world. Just as they told Heinrich Schliemann that Troy was a myth, imagine their surprise when he uncovered proof of Achilles, of Agamemnon and Paris and Odysseus! Yet did they acknowledge that they had been wrong? Did the scholarly world change its thinking? Galileo saw the universe as it truly was, and for this he was tried and found guilty of heresy. It is a miracle indeed that science has reached its current advanced state in the face of the resistance posed by "experts."

1.4

For here is the fact which the "experts" so vehemently deny: *They do not have all the answers.* How did man evolve? Every year scholars come up with a new explanation, contradicting even themselves. They argue endlessly at conferences and in thick books that nobody can understand, failing even to agree amongst themselves. How were the pyramids built? How were the monumental stone statues on Easter Island erected? How did the Maya build their pyramids and use advanced mathematics? We aren't entirely sure, scholars say. We have theories, but that is all.

But what use are theories if they do not give you final, solid answers? How can science claim anything if it itself admits it is not grounded in absolute certainty? Yet they always want more money for more grants to do their "research" so that next year they can tell us again that they don't have the final answer. Clearly something is wrong here. Clearly we must find better explanations for the mysterious phenomena taking place all around us. Clearly we must do better at seeking the Truth.

And the Truth is out there, if you only have the courage to look. That is the purpose of this book. *Doughnuts of the Gods* will require courage to read, courage to accept against the clamor of "scientific" orthodoxy, and no doubt it will be attacked and placed on that black list of books that are not to be read, that are to be hidden away from children. For Truth is real and sweet and sugary, not complicated and wordy and repetitious and wordy and repetitious, and the many holes in scientific explanations, the many uncertainties that scientists daily admit to having, can only mean one thing, even if they do not want to admit it: *They are Wrong, and therefore I, V.R.Y. Silly, must be Right.*

1.5

Yet if we put aside the prejudices of small, primitive minds, can we not look to the stars and wonder if we are indeed alone, or if out there, somewhere in the vastness, there are others like us, others who yearn to make contact? Since the scientific probability of there not being extraterrestrial intelligence elsewhere in the universe is virtually nil, should we not ask if such intelligences exist, and where they are? Yet where, then, the skeptics ask, are these others, these aliens? If they are there, why do they not come to Earth, recognizing familiar intelligence, and share their wisdom and knowledge with us? Why, when we turn our most powerful radio telescopes to the heavens, do we hear no answer? No, the critics say, there is no evidence, no sign of our extraterrestrial compatriots, and so they must not exist.

Silence means nothing.

But who amongst these skeptics has walked on another world? Did they not see when the brilliance of our science sent men to the moon in 1969? Where were they when the astronauts left footprints on another world? Let us ask rather another question: in a thousand years or ten thousand years, when our civilization is gone, what will visitors from another star think when they land on our moon and find the evidence, preserved for eternity in the endless vacuum of space, of men walking on the moon? What will they make of the flag, and of the machines left there? Will they too scoff at the notion that they are not alone, and will they not hunger to know us, to visit us, to talk with us?

Or have they already?

1.6

The human race is only about 250,000 years old. Before this our ancestors were little more than apes, evolving slowly toward *Homo Sapiens*, beginning more than 5 million years ago when we began to walk on two legs, distinguishing ourselves from the more primitive chimpanzees and gorillas. Our brains evolved slowly, first with *Australopithecus* and then *Homo Habilis*, followed by *Homo Erectus*. We spread out across the world, and yet when we look even at early *Homo Sapiens* we cannot help but be struck by how primitive he was; his tools, fashioned by the same sorts of hands that created the space shuttle, could only produce crude stone points and axes, chipped from flint and obsidian. Surely we are not like these men!

So what happened? Why are we so superior to those whose bones archaeologists uncover? Why, of the more than 200,000 years in which *Homo Sapiens* has existed, is it only in the past 5,000 years that mankind has grown civilized, capable of the kind of abstract thought required to create a computer or a jet aircraft? What must have inspired our ancestors to emerge from their caves and stand erect to strive for the stars?

It cannot have been evolution, for evolution is random. Something must have happened to accelerate our development. We must have had help, just as we have gone to the primitive peoples of the world and helped them.

1.7

How has this help occurred? Here we may rely upon the time-tested methodology of analogy. When we encountered the primitive peoples of our own world, what aid did we bring them? Technology, certainly; what impoverished nation has not benefited from the television, the automobile, the machine gun? And so we may reasonably surmise that when our extraterrestrial neighbors visited Earth, they brought our own ancestors such wonders as well. Erich von Däniken has noted that were extraterrestrials to come to Earth, they would be seen as gods by our own primitive ancestors, and surely this is right. The famous science fiction writer Arthur C.

Clarke has noted that once technology reaches a certain point, it is indistinguishable from magic. And so we may reasonably surmise that our own myths and legends may well contain hints of the "magic" of extraterrestrial technology in action, technology far beyond the capability of the primitive mind to comprehend, which then explains it as miraculous.

And yet what form did this encounter take? What, of all the wonders we have bestowed through our own encounters with primitive peoples, stands as the most important? And here, the answer once more is clear, and by analogy we can explain many of the mysteries of the ancient world, mysteries that have baffled archaeologists and historians. For what, above all else, do humans require to survive?

The answer is simple, as answers to complex problems so often are.

Food.

Two: Unexplained Mysteries of the Past

2.1

In the middle of the Sahara Desert in northern Niger there exists in the sand a perfect circle of stones. Archaeologists have no idea who built it or why, or even when. The archaeological report on the circle states that it is about 60 feet in diameter, and that "Roughly a mile away in each of the four cardinal directions, similarly crafted arrows point away from the circle, whose origin, purpose and age remain a puzzle."[1]

Who created this strange monument? Archaeologists will tell you it is a natural phenomenon, that the rocks were deposited by the motions of an ancient river, yet this is the middle of the Sahara Desert where no water flows. Or they will explain it as a holy place created by Bedouin tribesmen in the worship of some unnamed pagan god, or a way to calculate the solstice so they would know when to plant their crops. Theories in fact abound about such ancient monuments, each more fantastic and farfetched than the last, yet what each of these "experts" fails to do is look for the simple solution, the one right before their eyes, for to do so would be to reveal that the entire complex edifice behind which they hide is in fact like Potemkin's village, merely a front that they are afraid to challenge.

---

[1] *National Geographic*, March 1999, page 25; a photograph of the mysterious circle can be found at
http://photography.nationalgeographic.com/staticfiles/NGS/Shared/Stat icFiles/Photography/Images/POD/a/adrar-madet-massif-525421-lw.jpg).

2.2

Let us look at the circle again. What shape does it have? It is round, of course; circles usually are. Is this significant? And if we look at other unexplained monuments from the ancient world, monuments that, like our mysterious circle of the Sahara, belie explanation by current archaeological and historical theory, we see this pattern again and again: curves and circles. Stonehenge, which rises from the plains eight miles north of Salisbury in central England, consists of several circles of gigantic stones weighing 50 tons. Like the Sahara circle, archaeologists have no idea why it was built or by whom; it remains one of the great enduring mysteries of our distant past. They do know that the stones used to create it were cut with extreme precision from quarries 150 miles away, in what is now Wales (the Preseli Hills).

To cut such stones and move them so far; can we really expect to believe that stone-age men were capable of such a feat? To move even a single ton of rock (2000 lbs) requires all the technology we can muster today, a crane and an engine and a truck; imagine trying to move one weighting 50 times more (100,000 lbs!). And we are to expect that these same primitives could place such stones one atop another to form lintels? Even today, a quick visit to the site would dispel such notions in any reasonable person.

But there it is, undeniable! Archaeologists may want it to go away, but evidence is a truth that they cannot deny, try as they might! How did it get there, and more importantly, why was it built? That humans did not is obvious; an intelligence not our own must be responsible. So how do we, looking back through the mists of time, try to learn who this intelligence was?

2.3

And Stonehenge is not the only example of ancient circular constructions in England! A quick look at a map reveals such constructions all over the island, dating back thousands of years! Why would the ancient English have created them?

From whence did they come, and why? (Photo by the author)

The answer may be found if we remember the night sky. Just as we do, ancient men looked to the heavens to understand their place in the universe, and we see at Stonehenge and elsewhere that the monuments predict numerous astronomical phenomena, including equinoxes and eclipses. Yet why would primitive man need to do so? Some argue that this knowledge would help with farming, knowing when to plant and when to harvest, but surely this was not enough reason to move 50 ton blocks hundreds of miles. No, the real answer must lie in the realm of myth and belief, for it is only for these reasons that ancient man would exert such an effort.

A further clue has been overlooked by archaeologists, but can be found in the very data they have produced, that of the skeletal remains of the ancient Britons. An archaeological report states as follows: "Their relative lack of dental decay suggests a diet low in carbohydrates."[2] This would indicate that their diet was deficient, and it is only logical that when faced with such a predicament, people would seek out what they lack, in this case carbohydrates.

This leads us to the final solution to these ancient mysteries. Ancient man struggled to survive, to find enough to eat; this was true

---

[2] *National Geographic*, June 2008, page 37.

the world over. And so foods rich in calories would be most desirable to these primitives. It is simple logic to extrapolate a food that is both rich in calories and of the shape that Stonehenge and the Sahara monument both possess.

It was then, as I was walking down the street in New York, that the idea came to me in a flash of vision, an insight that like that of Newton when the apple fell on his head, one that was so profound that I was forced to sit down and gather my wits.

Doughnuts.

Three: In Search of Ancient Mysteries

3.1

It is one thing to make a profound discovery, but quite another to prove it with rigorous scientific research. Skeptics will say that Stonehenge and the Sahara circle are aberrations, exceptions, that their shape and astronomical alignments and the clear need for more carbohydrates in the ancient diet are mere coincidences. So if we are to proceed with ancient doughnut theory and convince those with open minds, we must seek out and explain new evidence from the ancient world. And as we will see, such evidence for ancient pastry visitations exists in abundance, once we are willing to cast off the shackles of scholarly orthodoxy and look at things objectively.

My search began, as all searches for truth and wisdom must begin, in the mysterious east, the Orient. Thus I boarded an airplane and flew to India, where after some fruitless searching I was directed to the great Maharabishi Ullabag Taturunos (at least I think that was his name) to whom I posed my burning scientific question: had ancient doughnuts in fact visited the Earth long ago, and had they left evidence of their passing in our legends and myths and ancient monuments? He regarded me curiously and asked in reply (for all great wise men answer questions with questions, as is well known), what I was doing in his house, if perhaps I was looking for the party being held in the nearby hotel, or maybe I was simply lost or had sustained a blow to the head.

But my quest for wisdom and knowledge was not to be so quickly dismissed! For I knew that wise men only ask such things to dissuade those inauthentic seekers who want only to hear platitudes that justify what they already believe. With some help from a $100 bill, the great Maharabishi's disposition became much more favorable and he entered a deep trance while I scribbled down notes furiously of what he said. Unfortunately, the only paper I had was a hotel napkin which later I was forced to use when I developed a runny nose, but I have been able to decipher most of my comments. He spoke as follows:

> Are you serious? Well, it's your money, I suppose: In distant times from long ago the most ancient forefather slumbered as eternity passed unheeded. And in this time the circle of existence was still. The Ah-natupa had not yet come into being. But in time the wheel of eternity began to move, and so spinning, the world was and became the first of mankind, those who resembled giant fish, crawled forth from the deep ocean deep and stood proud upon the land and looked around them to see what it was. And in time the Guzundisplat (I was unable to make sense of this part since later I set my cup of coffee on this section of the napkin) Atlantis emerged, the mother and mistress of all secrets and lost knowledge. Go to there, you who seek answers, and get out of here, since it is past my lunchtime and I need to take my daughter to soccer practice.

At this point I was escorted from the house and told not to return, a clear sign that I had touched upon a most great and ancient mystery whose initiates would do anything to protect. But as a civilized, Western man, I would not be so quickly dissuaded and so pursued the question of ancient pastries with ever more diligence.

3.2

Later, as I sat in the office of the chief of police awaiting deportation, I noticed the flag of India hanging behind his desk. For those of you who have never seen the flag of India, it is worth noting that in the center is a big circle. As they escorted me to the airport, I had time to think on this.

The Flag of India.

Another circle! And the great Maharabishi had mentioned circles in his trance. Clearly this could not be coincidence! There are circles everywhere in India, and elsewhere. It is one of the true universal symbols, going back thousands of years. In the Upanishads, the sacred texts of ancient India, we see the imagery of the circle used as the most basic element of the cosmos: "This vast universe is a wheel."

3.3

But what else had the Maharabishi said? Quickly I glanced at my napkin again, and the word leapt from the page: Atlantis! The lost continent! The source of ancient wisdom! If ancient doughnuts had visited the Earth and were responsible for the myriad ancient monuments far too complex to be created by primitive man, then surely the best evidence would lie in the legendary sunken continent! Quickly I consulted a map of the Atlantic, but found no sign of Atlantis save for the Azores; clearly they had once been the high mountains of that lost and forgotten land. Fortunately, scientists have mapped the bottom of the Atlantic Ocean using advanced technology, and so it was to one of these maps that I then proceeded.

Sadly, this map showed nothing. How could the geologists have missed something as obvious as a sunken continent? Clearly it was academic self-deception, so common in the Ivory Towers of our universities and museums. How can they claim to know what is on the bottom of the ocean, after all? Fortunately, others have not been so quick to relegate Atlantis to fiction.

The earliest source for the Atlantis story is the philosopher Plato, and yet he himself tells us that the story did not originate from him but rather from a relative of his great-grandfather, a wise man named Solon. Solon in turn had learned of Atlantis from Egyptian priests, and even the least educated among us know that the Egyptians have the oldest civilization in history. If records of even earlier civilizations exist, it will be with them, and as we will see later, the evidence for ancient pastry contacts with Egypt is overwhelming.

Solon recounts the story of a great war fought between Atlantis and the city of Athens 9000 years before his own time. Since both Solon and Plato lived about 2500 years ago, this would place this war at about 11,500 years ago. Athens was victorious in the war, but in his description of it Plato tells us a great deal about Atlantis. It was an island "larger than Libya and Asia together," which would make it the largest continent in the world. It is possible in fact that North and South America are themselves remnants of Atlantis, given its vast size, which would indicate that only the eastern part sank into the sea and we may look for evidence of Atlantis among the many mysterious cultures of the New World, including the Maya and the Incas, about whom we know next to nothing. It is perfectly logical that from the point of view of the Egyptians and Greeks, the sinking of the eastern half of the continent of Atlantis would seem like the entire continent had gone underwater, given the vast size of the Atlantic Ocean that resulted from the cataclysm.

3.4

But the question remains: can we link Atlantis with ancient doughnuts? For surely the Atlanteans, like the builders of Stonehenge, would have had contact with such sugar-loaded alien visitors, and indeed, where else could the Atlanteans have gotten their high culture if not from such glazed and powdered and jelly-filled extraterrestrials? Fortunately, Plato's record of Atlantis answers this question conclusively. He tells us that Atlantis was the share of Poseidon, the ocean god of the Greeks, and we should note that the name "Poseidon" contains two occurrences of the letter "o", or omicron—the Greek rendering is *Ποσειδῶν*, and in the even more ancient Linear B tablets, the name is rendered *Po-se-da-o*. The letter omicron, of course, is another example of circularity, and anyone can see that it resembles a doughnut: O. To argue that this is coincidence is to stretch credulity too far. No one can deny that doughnuts are round!

3.5

But let us now think further. Yes, Atlantis lay in the Atlantic, and it was destroyed by a great cataclysm. But what other sources might there be for Plato's story? There is one that clearly shows the influence of ancient doughnuts, one which I am surprised has not been seen before.

In the middle of the Aegean Sea there is the island of Santorini. Geologists have proven that this island was once whole, and we know that ancient and mysterious peoples lived there long ago. But between about 1600 and 1645 BC, the center of the island exploded with a force of many nuclear bombs, wiping out the civilization whose remnants now beckon to us. We can only guess at how many people died; the ruins tell us only that they were there.

What could cause such a blast, which is beyond even our most powerful bombs today? A volcano? I have visited Hawaii, and the volcanoes there may bubble, but explosions are limited. Volcanoes *build* islands, you see; they are part of the creative force of continental drift. No, what destroyed Santorini must have been something else, something far more powerful and intentional.

Had the primitive peoples of the island offended the "gods"? Had they committed some affront and angered them?

We may never know.

But we do know this: There are ruins on Santorini, signs of the terrible devastation. And if we look closely enough, we may even see evidence of who was responsible.

For what remains of the island resembles nothing so much as a partially eaten doughnut! Is it a message from the doughnuts? A warning? Did the people of Santorini offend their pastry overlords? Certainly by Plato's time the destruction of Santorini would have still been in living memory.

And yet scholars, trapped in their Ivory Towers, scoff at what is so obvious.

Doughnuts destroyed Atlantis, and they destroyed Thera.

3.6

Plato's account includes a description of Atlantis, most particularly the great city there. This city, we are told, was made up of an inner and outer city, connected to the sea by canals. Most importantly, however, these cities, and the walls around them, were circular!

Surely, you might say, a city might be circular for other reasons, but consider this: the circular dimensions of a city would only be distinguishable from the air, and we know that ancient men had no airplanes or spacecraft. Why then build this way? Clearly the circular dimensions of Atlantis' cities were meant to send a signal skyward, to the gods. Was it an invitation, a plea for more delicious sugar and the high number of calories that came with it? And did the Atlanteans eat too much sugar and offend the extraterrestrial pastries, causing them to use their tremendous caloric power to sink the continent under the waves? We can only speculate, but perhaps someday oceanographers will plumb the depths of the Atlantic and discover an Atlantean archive that will tell us the true story of this fascinating lost continent and its relation to intergalactic doughnuts.

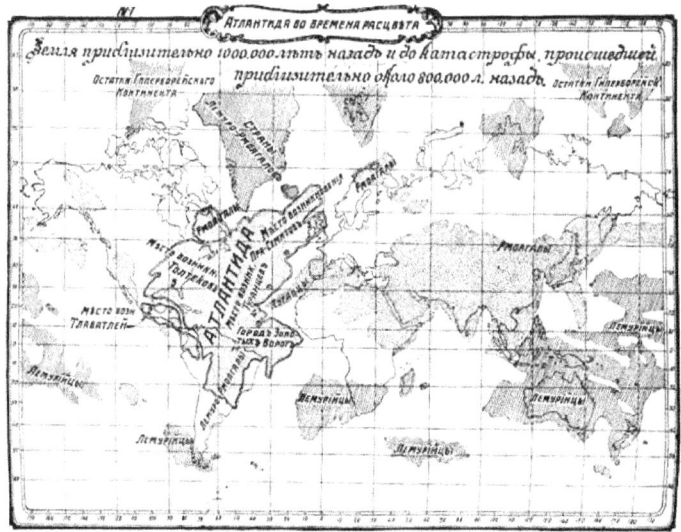

Map of Atlantis
(https://commons.wikimedia.org/wiki/File:Map_of_Atlantis.jpg)

3.7

But has some of the Atlantean wisdom survived? As I mentioned before, it is likely that North and South America are the remnants of Atlantis, and so some Atlantean antiquities may still be there. With this thought in mind I made my way to Ecuador and to the mysterious cave called Cueva de los Pendejos, well known to contain ancient mysteries. Climbing to the top of a hill with my guide, we walked into the open cave entrance and were soon enveloped in total darkness. Pulling out a set of flashlights, we made good time through the chambers, the weight of antiquity weighing heavily on us. I could not help but wonder: had the doughnuts who once came here come from other stars? What could they teach us? What wisdom of the ages might they impart? I asked my guide, Juan Juarez, if he had ever had any strange experiences here.

"Si, Señor," he told me. "There are many strange things in the Cueva de los Pendejos."

"Take me there," I instructed.

Deeper and deeper we descended into the hill, until at last Juarez stopped and turned to me. "From here you may only go if you are ready," he said.

Breathlessly I answered.

"I am."

"I mean, do you have to use the toilet? There are no facilities within, Señor."

I tensed. Truly we were entering the unknown.

"I went at the hotel."

He regarded me in the low light of our lanterns, his face inscrutable, and for a moment I could see his Atlantean heritage there, buried, yes, in the thousands of years since that noble continent vanished beneath the waves, but his ancestors had survived, and somehow they had passed down their wisdom to him despite the destruction of his culture by modern man. And as he nodded I saw a kindred spirit, one who like me sought the wisdom of ancient pastries, of jelly-filled knowledge, of chocolate frosting on a sugary, circular base.

"Follow," he said.

3.8

We entered a room filled with wonders. Golden artifacts littered the floor and the ancient stone shelves, and among these were tablets, metal tablets whose surfaces were covered with ancient script. Perhaps this was the language of Senzar, and some of these tablets held the legendary *Book of Dzyan!* Quickly I pulled out my camera, but Juan Juarez stopped me with an outstretched hand.

"Such mysteries, Señor," he intoned, "must not be photographed."

I stopped in place; his voice had a commanding tone that had not been present above ground. I also realized that I had forgotten to bring a flash, and so lowered my camera and reached for my notebook.

"Is it permissible to make a drawing?"

"Si."

Picking his nose, Juan Juarez seemed lost in thought as I brought my flashlight to one of the tablets and began to sketch what I saw. And as I did, it seemed that the meanings of the words were becoming clear, as though the Atlanteans who had placed these records here had been waiting, not for me, but for humanity, to crawl out of our primitive past and look at the stars as they had, as they had been guided to by the extraterrestrial doughnuts from long ago. Perhaps now, for the first time in eons, we are ready to face the truth of our own mysterious past.

I had little time to copy, however, since our light began to grow dim, and with my notebook clutched firmly in hand—for to lose it would be to lose the irreplaceable knowledge it held—I ascended from the deep cavern with my guide, there to stand under the bright light of the sun once again, changed forever by what I had seen.

Atlantean Writing

Four: Myths of the Ancients

4.1

One merely needs eyes to see and read, and ears to hear and listen, and most of all a mind open to what these senses tell it, unburdened by preconceptions and prejudices fed to it by the powers of orthodoxy. For as I emerged from the mysterious caves and set to translating the text I had copied, it began to influence me in deep, intuitive ways, opening my eyes and ears to reality. For is it not true that the signs of prehistoric pastries are everywhere? Do we not see them every day, and yet pay little attention as we hurry through our mundane lives?

How else can the vast body of evidence I will show be explained?

Let us begin at the beginning, a very good place to begin.

Early man looked to the heavens and drew what he saw; these images have been preserved for us worldwide in caves and on rocks, and in all of these the same feature appears: the images of men are presented with circles around their heads. Can this all be coincidence?

Imagine for a moment a primitive savage, huddled against the cold of night. And then a giant doughnut descends from the sky. Our primitive friend sees that it is round, and that it is sweet; in his half-starved condition how could he *not* think this was a great sky god? Naturally he goes and tells his friends, and they ask him to draw what he saw. Over time these drawings, and the accounts of our friend and those like him who also encountered these strange and wondrous pastries, become the stuff of legend, passed down around campfires.

And in time, we may surmise, these extraterrestrial doughnuts begin to help the savages, for the intergalactic pastries realize that here is a species that may one day reach the stars themselves, if they are only helped along. And so they breed with these primitives, producing sweet hybrids, and they share with them some of their knowledge and wisdom. In time the savages elect a king, one whose duty it is to interact with the doughnut "gods," and civilization is born.

Fanciful? Farfetched? Insane? The scholars in their Ivory Towers, cut off completely from the modern world, will no doubt say so. Show me the evidence! they will cry.

That's not how science works!

It's nothing like anything we have concluded!

My colleagues will laugh at me if I say that!

4.2

Crucial to civilization, of course, is the wheel. Without it, we are no different from our ignorant and savage forebears. But the wheel is more than simply a tool; it is clearly a symbol. It is round, and it has a hole in the center. Nothing could be more doughnut shaped than this! We know that primitive man had no wheel, so where did it come from? The only possible, logical conclusion is that it was introduced, and that it reflects those who introduced it.

Giant intergalactic doughnuts.

4.3

Wheels, of course, appear in more than just archaeological sites; the legends of our forebears are filled with them. In the biblical book of Ezekiel, the prophet is brought before God, who rides a particular vehicle. Ezekiel tells us:

> Now as I looked at the living creatures, I saw a wheel upon the earth beside the living creatures, one for the each of the four of them. As for the appearance of the wheels and their construction: their appearance was like the gleaming of a

chrysolite; and the four had the same likeness, their construction being as it were a wheel within a wheel. When they went, they went in any of the four directions without turning as they went. The four wheels had rims and they had spokes, and the rims were full of eyes round about. And when the living creatures went, the wheels were beside them; and when the living creatures rose from the earth, the wheels rose. Wherever the spirit would go, they went, and the wheels rose along with them; for the spirit of the living creatures was in the wheels. When those went, these went; and when those stood, these stood; and when those rose from the earth, the wheels rose along with them; for the spirit of the living creatures was in the wheels. (1:15-21)

Not even in a Dunkin' Doughnuts will you find a more accurate description of a doughnut. Further, Ezekiel, who we must remember, is actually viewing an intergalactic doughnut, clearly associates the doughnuts with wheels, which he must have known were vitally important to his civilization.

4.4

Other examples of ancient doughnuts abound. If you visit Rome, you cannot help but see the famous Coliseum, but have you ever looked at it closely? From the ground it looks like a giant stadium, and archaeologists and historians will tell you that this is just what it is. But from above, from a vantage point that it was impossible for the ancient Romans to achieve, the Coliseum is circular!

It is also a vast monument, and so we must ask ourselves: just as with the pyramids of Egypt, how could primitive man have constructed such a thing? And why would they make it round, if you could only see this from the air? Clearly there must have been some other important purpose to this structure, especially since it is in the middle of the city.

Of course, if you are a giant doughnut, the function of the Coliseum is clear: if you fill it with powdered sugar, you can lower the doughnut into the space and get your sugar covering more easily, a fact I recognized while waiting for my snack at Krispy-Kreme. And yet no archaeologist I asked had thought to examine the ruins for traces of powdered sugar! How can they overlook something so obvious?

The Sweetness that was Rome (Photo by the Author)

4.5

In 1513 the Turkish admiral Piri Reis created a map of the Atlantic Ocean that shows with perfect accuracy the coasts of Africa, South America, and Antarctica. And yet we are told by historians and archaeologists that many of these places had not yet been discovered then! How can this be? Antarctica is covered in ice, and yet Piri Reis depicts its coastline with complete accuracy; likewise South America is covered in jungle, impenetrable to the human eye! Further, Piri Reis tells us that his map was based on even earlier maps, dating back to the time of Alexander the Great!

So we must ask, was the Antarctic free of ice in 300 BC?

Was South America a desert?

Remember that we have established that South America was once part of Atlantis. Its jagged coastline is the remnant of the great catastrophe that felled that once-mighty civilization, and this no doubt was what made its coastline jungle-free and therefore able to be mapped. But what of Antarctica? Clearly the original mapmakers must have seen its coast without ice, and this can lead us to only one conclusion: it must have been much further north—thousands of miles further north—in 300 BC.

Logic dictates, therefore, that between the sinking of Atlantis (or the eastern part, in any event) and today, the entire continent of Antarctica must have been connected to South America and that both must have been closer to the equator. We must therefore conclude that a second catastrophe, following the sinking of Atlantis, must have happened after 300 BC.

4.6

But the true value of the Piri Reis map has been overlooked by even such scientists as Charles Hapgood, Erich von Däniken, and Graham Hancock, who miss the clear source of the map, presented by Piri Reis himself. There, directly in the center, more prominent than the coastlines, are two giant doughnuts! A third, smaller one, lies in the northern part of the map, and a fourth in the southern part, and yet a fifth in the very center! Further, from each of these doughnuts extend straight lines, reaching downward toward the Earth.

Could primitive man have made such straight lines, extending over thousands of miles?

What possible purpose could a Turkish admiral have had in depicting such features out over the open ocean?

Clearly the explanations of the historians and archaeologists are wrong, and we must have the courage to rethink this fascinating ancient artifact.

Intergalactic doughnuts clearly created the Piri Reis map, and now for the first time we have actual pictures of them! But what of the lines? What significance do they hold? Is there evidence of them elsewhere?

In fact there is.

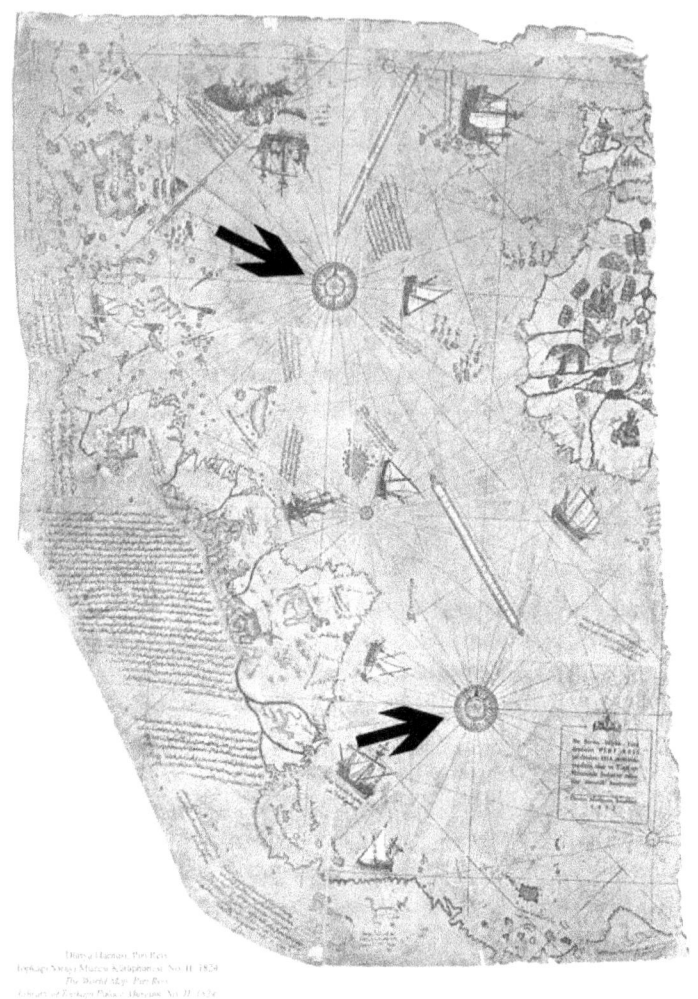

The Piri Reis Map. Note the doughnuts.
(https://en.wikipedia.org/wiki/Piri_Reis_map#/media/File:Piri_re
is_world_map_01.jpg)

In ancient Egypt there was a pharaoh named Akhenaten, who reigned between 1353-1336 BC. Early in his reign he had a religious epiphany, and he directed the Egyptians to abandon their old gods and worship a single god in their stead, called the "Aten." This would all be unimportant, since there were plenty of pharaohs in ancient Egypt, except for two facts. First, Akhenaten had images of himself made, and it is clear from these images that he looked nothing like the other Egyptians. His head is elongated and his body is rounded in a way that no male body should be; his hips are those of a woman. Why is this? Why would the pharaoh, the king, want to depict himself in such a strange way?

There can be only one answer: Akenaten was either an alien, or an alien-human hybrid. Consider the evidence: aliens have peculiarly shaped skulls, as did Akhenaten. They have big, slanty eyes; so does Akhenaten. Doughnuts are rounded, and so were Akhenaten's hips. And aliens do strange things; what could be stranger than this pharaoh?

Akhenaten
(https://en.wikipedia.org/
wiki/File:Akhenaten.jpg)

Space Alien
(Photo by the Author)

But if Akhenaten was an alien, who was his extraterrestrial father? The answer, as it turns out, is right before us, just as it was in the Piri Reis map. It is the Aten itself! Note how the Aten is round, like a doughnut! Note the straight lines descending to earth, just as they do in the Piri Reis map, reaching downward to help mankind. Akenaten was clearly sent by the intergalactic pastries to reform Egypt, which must have strayed from their teachings.

Based on these incontrovertible pieces of evidence, we can therefore surmise the true purpose of the Piri Reis map. Atlantis had sunk beneath the waves; who knows how many lives were lost, what shock and trauma the survivors must have felt? Immanuel Velikovsky, a psychiatrist by training, tells us that ancient man, having endured these horrific catastrophes, underwent a case of specieswide amnesia to blot out the terrible memories.[1] But such widespread amnesia could not have been the result of merely the primitive mind; it must have had help. What the Piri Reis map clearly shows is that extraterrestrial doughnuts, following the cataclysm of the sinking of Atlantis, came to our aid and blanked out our memories with psychotronomatic rays of some sort, an event recorded in the sources that eventually found their way into the hands of admiral Piri Reis.

Thus, it is clear that Akhenaten was bred and sent by intergalactic pastries to restore mankind to the knowledge of extraterrestrial doughnuts. But sadly for all of us, his efforts were stymied by the ignorance of the Egyptians, whose religion, though originally reflecting doughnut theology, had backslid into the kind of mind-numbing babble pushed by people who expect you to believe the most preposterous ideas imaginable, like the notion that extraterrestrial enchiladas are responsible for ancient mysteries.

---

[1] See his book *Worlds in Collision,* pp. 302-315. And note also that I'm citing a scholarly reference. So much for *your* complaints, Dr. Vedeler!

4.7

Deep in the jungles of Central America, the civilization of the Maya emerges suddenly from nowhere. Pyramids appear in sites like Chichen Itza, Tikal, and Palenque, giant cities rising from the dense forest (perhaps as rebel bases against some malevolent galactic empire?) greet the traveler and call to him with their mystery. Where did they come from? Who built them, and why? They do not fit the convenient theories of archaeologists and historians, who would have you believe that the ancestors of the local peoples created them. Yet as we are taught in our outstanding American system of public education, when the Spanish under Christopher Columbus discovered the Americas in 1492, they found only primitives living amongst the ruins.

How could such peoples have built these great monuments?

Yet it is true that the monuments of the Maya, and those of neighboring cultures such as Teotihuacan and the mound builders of North America, show images of people, so we know that whoever the builders of these mighty cities were, they allowed the locals to come in. Is it so preposterous that extraterrestrial doughnuts, which we have seen such abundant proof for elsewhere, might have raised these great cities as a sort of intergalactic "peace corps" to aid the savage and ignorant humans below?

The largest of the two great pyramids at Teotihuacan, which have been dated by local legend to a time when human beings had not even yet evolved, measures 223 meters by 111 meters, far beyond the capacity of even modern man to construct. They are surrounded by myriad other buildings which, seen from the air (recall the Roman Coliseum), form an exact map of the Milky Way galaxy, which you will note, when seen from above, is round, like a doughnut! And how is it possible that the famous Aztec calendar, which predicts the future with uncanny accuracy, is shaped like a doughnut, no different than the description of God's chariot in the first book of Ezekiel?

4.8

But let us return to the Maya. At the site of Palenque in Mexico, a pyramid was discovered that was filled with inscriptions. Surely these would give us the answers to our questions about the past, if we could only read them! Fortunately, scholars have made great progress in deciphering the ancient Maya language, discovering legends such as that of Queen Moo and the brothers Coh and Aac, and how she was forced to flee to Atlantis after her lover Coh was killed, proceeding from there to Egypt. This connection to Atlantis cannot be coincidence, since I have already established that Central America was once part of that lost continent. Clearly the Maya are related to the Atlanteans, just as my South American guide Juan Juarez carries their noble blood.

The pyramid of the inscriptions not only holds answers in the form of its glyphs, but has also produced one of the most important archaeological finds of all time, definitive proof of ancient extraterrestrial visitations to Earth, which has been verified by numerous world-famous scholars. I refer, of course, to the sarcophagus lid of the extraterrestrial named Pacal, which clearly shows a rocket in flight and a man piloting it. But there is much more to the image than this, of only we take the time to look.

For there, above the rocket and clearly its destination, are two doughnuts!

Look at them closely. They bear faces, and around their edges are smaller circles, much like those of the Aztec calendar. These must be the "eyes" referred to by Ezekiel. And on Pacal's lap there is another doughnut, clearly his means of communicating with the "gods" who are calling him into the heavens for a sweet and tasty snack.

There is more. When the lid was raised a body was found below, and this body wore a mask that clearly does not show a human face. It is made of jade, and is blue, like the ancient gods of India! Coincidence? The archaeologists and historians would have you believe so. But how, save for the intervention of extraterrestrial doughnuts, could this color be found in both places, thousands of miles apart?

Pacal was clearly a human-doughnut hybrid, like Akhenaten of Egypt, and when we are able to read the inscriptions they will unquestionably verify this. I approached the archaeology department of the national museum of Mexico and urged them to immediately do a scientific analysis of the bones to find doughnut DNA, but they told me I was an idiot and to get out of their office. Clearly there is a scholarly conspiracy that is trying to keep the Truth hidden.

The Pyramid of Pacal (Photo by the Author)

Academic spy at Palenque (Photo by Georg Vedeler)

Pacal's Sugary Ascent
(By User:Madman2001 - Made it myself based on several drawings,
CC BY-SA 2.0,
https://commons.wikimedia.org/w/index.php?curid=4998025
https://en.wikipedia.org/wiki/K'inich_Janaab'_Pakal#/media/File:
Pacal_the_Great_tomb_lid.svg)

4.9

But conspiracies cannot deter those with truly open minds. The answer lay in the past, in the hints left to us by the ancients, in their monuments and words, and so I continued my quest in museums, dusty with antiquity. In each there were, if you opened your eyes and your mind, images of doughnuts, images of our sugary past, ignored by the arrogant academics. But where was the final connection, the final piece of the ancient puzzle? Where would I find that last, vital clue that would reveal our ancient origins?

Those who would challenge the scholarly orthodoxies, the stodgy professors in their Ivory Towers, are so often ridiculed, but it is they who propel science forward, who lead to the great discoveries. Struggling in obscurity, the great minds are those who are patient, who do not merely accept the easy, simple explanation of popular scholarly theory, but who are willing to look beyond it, who are willing to let their minds roam free and see what so often is right in front of us. And it is these thinkers, these free spirits, who also can see when the clues are placed right in front of them, placed before millions and get ignored, until one open mind spies the truth.

For there it was, as plain as day. The final piece of the puzzle, the final clue of my quest, the thing that would guide me to the answer of our ancient past. A photograph from Paris, from one of the great museums of the world.

The Louvre. A pyramid!

Hurriedly I packed my bags for France.

Five: The Ancient Marvels of the Pyramids

5.1

Paris is a beautiful city, and it was laid out in all its splendor as my flight circled to land. Eagerly I passed through customs and found a cab, which took some work since for some reason the driver didn't have the courtesy to speak English. (As a famous philosopher once said, "It's like those French have a different word for *everything!*") But soon, after a quick lunch at the local McDonald's, I was at the Louvre, and eagerly I made my way into the galleries. The Louvre is an amazing place, and as I walked through its many rooms, I could not help but wonder at the ancients who had built it.

Who were they, and how had they created such an architectural wonder?

How had they known about such modern building techniques hundreds of years ago? What secret messages might be hidden in its architecture?

The past is filled with mysteries such as these!

Perhaps someday a scholar will look into this, for surely this amazing building must have been here when the first primitive Frenchmen arrived in their berets with their baguettes, singing "Frere Jacques" and drinking their famous wine. Perhaps these mysterious builders were also the ones to teach them how to make brie!

But such speculations, I realized, must wait for another day, for I was in pursuit of ancient knowledge about pastries, not cheese. I entered halls that were filled with more ancient treasures, and then, before me, I found the answer I sought.

5.2

Standing at over ten feet tall, the law code of the ancient Babylonian king Hammurabi is, to say the least, an imposing monument, all the more so since it is made of black diorite and there is no stone in Mesopotamia. So where did the law code come from? Are we to believe that the primitive Babylonians, without the benefit of iron, somehow carried this massive monument to Mesopotamia and then carved such intricate and beautiful designs? Diorite is an extremely hard rock, and so carving it, difficult enough today, would have been impossible for the bronze-age Babylonians. And further, even though Assyriologists claim that the many signs on the code are cuneiform writing, a cursory look at the code and an actual cuneiform tablet shows that the symbols are completely different, and oriented in a completely different way.

As I stood and stared at this magnificent artifact, my intuition told me that the "experts" must be wrong.

But if so, then what did the mysterious symbols mean? What message from the most distant past might they convey?

Let us look further, releasing ourselves from the biases we are taught in school, from the reliance on those who think they know so much just because they have spent decades of their lives in careful study of these things, just because they have studied volumes of evidence and learned ancient languages. Let us rely instead on our first impressions, on our untrained instincts, on what we already know must be true even before we set eyes on any evidence!

Note the top of the stela. Those who tell us that the symbols are Hammurabi's code never discuss this, but it is a picture of Hammurabi standing before a god, picking his nose in respect. And if we look even more closely, we see that this god is handing the Babylonian king a short stick and what is clearly a doughnut!

A doughnut of the gods!

Quickly I began to look at other Mesopotamian stelae, and there, on many of them, was the same scene! This cannot be coincidence, as the Assyriologists will tell you. It cannot be without meaning. The "gods" were handing doughnuts to the kings! I glanced at the notes

I had made in the Cueva de los Pendejos, and there too were these symbols. Like the Egyptians, the Mesopotamians were old, old enough to remember Atlantis.

And old enough to leave us their images of the ancient "gods."

Hammurabi's Stela; note the doughnut in the hand of the god. (Photo by the Author)

5.3

History is not a subject to be taken lightly. It requires a talent for deduction, for seeing patterns in the faint traces left to us by those who have gone before. It requires a talent for interpretation, and for consulting and referencing previous scholarship. An important historian once put it well, and we should pay heed to his words: "History is not a subject to be taken lightly. It requires a talent for deduction, for seeing patterns in the faint traces left to us by those who have gone before. It requires a talent for interpretation, and for consulting and referencing previous scholarship. An important historian once put it well…."

I made certain to take photographs of the stelae I had examined, adding to the vast body of evidence I had gathered in my travels. And I realized that only one question remained: where had the intergalactic doughnuts landed? How had they interacted with our primitive ancestors? The mysterious writing on Hammurabi's stela, like the mysterious writing in Pacal's temple at Palenque, no doubt would answer this question, if only we could read it. But I seemed to be at a dead end, since that knowledge still eluded us, and would no doubt only come after we had discovered the doughnuts' ancient landing site. Invigorated by my new discoveries, I made my way out of the museum.

And there, before me, was the answer. The ancients who had built the Louvre had left the final clue in plain sight!

A giant glass pyramid stood glinting in the late afternoon sun.

The Louvre Pyramid. Clearly a sign! (Photo by the Author)

5.4

The god, handing the symbol of the doughnut to the king, in a stela housed in a museum with a pyramid. Clearly this could not all be coincidence! The pyramids are important, but how? This question troubled me, for of all the shapes I had encountered in my travels, these could not be in any way considered round.

How do we explain such ancient mysteries?

Like the mysterious civilizations of Central America, the monuments of Egypt sprang from nothing along the banks of the Nile, or so the archaeologists and other experts claim. Suddenly—poof!— pyramids appear, so large we could not even imagine building them today, even with all our modern technology. And even to the untrained eye we can see that they are like the pyramids of the New World, which also appear from nothing, or so we are told.

But inventions do not come from nowhere. They must have an origin, and spread from there. This is common knowledge. So we know that a pyramid in Egypt and a pyramid in Mexico or Peru or Southeast Asia must derive from a common source, a common culture. We can see evidence of this in the many shared words between languages of the Old World and the New. Atlantis was key, but was it the *source* of ancient civilization, or merely a conduit? Certainly there is too much evidence of influence on Atlantis by doughnuts to be dismissed, but the doughnuts must be much older than the Atlanteans, and so if we are to find their original landing site, we must look to monuments that predate Mesopotamia, that predate the pharaohs of Egypt, that predate all known history.

And if we look closely, we see that such a monument exists, there on the banks of the Nile, older than Egypt itself.

5.5

The Sphinx has long intrigued travelers with its mysterious stare, as if challenging us to understand its great secrets. Traditional Egyptologists tell us that it was built by the pharaoh Khafre in the middle part of the third millennium BC, giving as evidence several texts, but there is far from universal agreement on this. And if the "experts" cannot give us a conclusive answer to a question, there is no reason to believe anything they say. History is too important a subject to be left to those without firm answers to all questions!

Fortunately, other scholars have weighed in with more certain conclusions. The Yale-educated geologist Robert Schoch has proposed solid evidence that the weathering of the Sphinx clearly indicates that it was carved long before Khafre, as far back as 7000 BC. And since geology, like mathematics, cannot lie, his proposition that this indicates that an ancient culture once existed to do this must be correct; there can be no other alternative explanation.

But why carve such a sculpture? The Sphinx is clearly not human, but has the face of a human. The body of a lion? We have already seen how doughnuts mated with humans to produce such hybrids as Akhenaten and Pacal (and probably Hammurabi), but this does not answer the question of where humans came from in the first place. Archaeologists will tell you that we evolved from simpler apes, but this does not explain where human intelligence came from, why suddenly our ancestors went from being scarcely more than animals to the highly intelligent creatures we are today, capable of travelling to the moon and beyond.

Is it possible that even before the later hybrids that alien pastries were manipulating our DNA?

What more noble a creature is there than the lion, the king of the jungle? Is it so preposterous that ancient extraterrestrial doughnuts mixed lion DNA with that of our proto-human ancestors, providing the last push of energy that turned us from monkeys into the lionlike species we are today?

As I walked around the Sphinx, I could not help but wonder at the many mysteries of our most ancient past.

5.6

Soon, however, my attention was drawn to the pyramids themselves. Clearly the ancient builders of the Louvre had directed me here; clearly this was the final clue. The sky was blue overhead, the air sharp and dry, and around me the camels bleated their protests as they ferried tourists around these ancient monuments. Did any of these people have any idea what it was they saw? Imagine, if you will, how these monuments must have been built, the amazing effort that must have gone into them. Egyptologists assure us that it was the Egyptians themselves who did it, but in the same breath we are informed that there are 2,500,000 blocks in the Great Pyramid alone, all cut to exacting fits that even today you cannot pass a razor-edge through. And this from a culture that had barely mastered copper tools?

Clearly there is something wrong with the stories we are told by the "experts."

And yet there the pyramids are, appearing suddenly from nothing, in a place without culture before, just a dry desert that no one could have ever inhabited; even today no Egyptians live far from the Nile. Do those with such easy explanations therefore explain why the height of the Great Pyramid, multiplied by 24,468, equals the radius of Mars? Can this be mere coincidence, especially considering the fact that there are circular formations all over Mars that are even now being explored by NASA rovers?

And then there is the fact that the pyramids are precisely aligned with the constellation Orion, in a formation that can only be viewed from above, which clearly indicates that they are far older than the dates given to us by archaeologists, who rely on things like carbon-14 dating that cannot even give us an exact result! At the site of Tiahuanaco in South America, astronomical alignments have also dated that site as far older than the date given by archaeologists. Why is this?

Simple. Carbon-14 dating cannot date stone. So which theory is really more likely, that of uncertain archaeologists or the inerrant mathematics of astronomers?

5.7

Only extraterrestrial doughnuts can explain these things.

The skeptics will say, of course, that we are misinterpreting, and of course, as a scientist, I will be the first to admit my error if ancient doughnut theory can be shown to be wrong. But there are too many things that the scholars have not debunked, too many things that are unexplained. And even the brilliant astronomer Carl Sagan tells us that ancient aliens may have visited Earth in the past, after all.

5.8

But why? I looked up, at the heavens, and then down, at the dirt beneath my feet.

And as with so many things, the answer was there, so obvious that scholars, trapped in the rigid preconceptions of their self-assured orthodoxy, could never see.

For the pyramids are surrounded by dirt.

A little research, done later, merely affirmed my observation: the entire Earth, even under the oceans, is covered in dirt. And if there is one thing that is clear in this chaotic world, it is this single, absolute scientific fact: if you are eating a pastry and you drop it on the ground, it will get dirt on it.

Sugar glaze is sticky.

So ancient extraterrestrial doughnuts came to Earth, perhaps millions of years ago. They were a survey team, in all likelihood, and they saw that this planet was rich in the essential ingredients for intelligent life: grains to produce wheat flour, yeast to make it rise in the oven, and cane sugar to sweeten it. As well, they encountered a species of primitive apes that could be bred into something new, something special, if only they were given the proper guidance. And so the doughnuts came back again and again, performing their mysterious experiments, until the ape-men had learned to appreciate sweetness and were ready for the next step in the cosmic pastry plan.

These men, these first men, were then mated with lions, their DNA mixed in careful bakeries, which gave them the firmness, the aggression, that they would need to fully serve the doughnut plan, and it was at this time that the interstellar pastries scattered them across the Earth and taught them how to grow crops, how to make the flour and the yeast and the sugar that were so essential to the movement of doughnuts across the cosmos.

But there was a problem. The Earth, so suitable in so many ways, was dirty. The grit of dirt tormented the doughnuts, sticking to their sugar glaze, until at last the decision was made: if humanity, their grand experiment, was to succeed, if we were to achieve our true destiny in the heavens, some means had to be devised to protect the extraterrestrial doughnuts from the dirt of our world. And so worldwide, from Mesoamerica to Southeast Asia to Egypt, the ancient doughnuts began to use their superior technology, technology far surpassing anything we have today, to build pyramids upon which they could land and spread their sugary goodness without the risk of contamination by our gritty, dirty soil.

The relationship between the pyramids and doughnuts is so clear, therefore, and it is only scholarly prejudice, the stubborn unwillingness to let go of old ideas indoctrinated in the classrooms of fixed and rigid minds, that prevents archaeologists and historians from looking at the past objectively and seeing the great destiny it holds for all of us.

For this is the final message of the intergalactic doughnuts: the future will arrive in the future, and it will affect the rest of our lives, for that is where we will be in the future. Someday the interstellar pastries will return, and on that day the true meaning of our lives will become clear to even the most stubborn skeptic. Sweetness will replace strife, war will end, and under the direction of our extraterrestrial pastry guides, a new period of human consciousness will emerge, with jelly filling and powdered white coverings, chocolate and sprinkles. And if you do not listen to V.R.Y. Silly, you will face the *hypoglycemia of the gods!*

The truth none dare admit! (Photo by the Author)

Commentary

1.1

This sort of introduction is common in the presentation of pseudoscientific theories. Rather than stating the problem they intend to solve, the pseudoscientist appeals to the emotions of the reader, trying to hook them with a romanticized setting. Good examples include the first chapters of Immanuel Velikovsky's *Worlds in Collision* and Erich von Däniken's *Chariots of the Gods?*. Of course, this is not limited to pseudoscience, appearing in both popular nonfiction and fiction, and so while we can hardly criticize authors for using it to generate sales, it does immediately distinguish their work from the typical scholarly monograph.

More seriously, many pseudoscientific works begin with broad statements that do not even lay out their thesis. Following these statements they proceed directly to the presentation of evidence, never telling us by what methods they are evaluating it. The impression one gets is that they are doing this evaluation ad hoc. As a result each broad area of pseudoscience tends to be made up of a number of smaller pseudoscientific theories that are not mutually compatible; Christian creationists have a different theory on the origin of life than Hindu creationists, for example. Because of their paradigm rigidity, pseudosciences are unable to merge with each other and form complete and normal scientific paradigms which would advance the study of the field. Instead, there is a pseudoscience for every taste, every opinion, and no consensus among pseudoscientists themselves is ever achieved. Customer satisfaction becomes more important than facts.

For the mysteries of the ancient world, this trend is widespread. We have theories about lost continents and other vanished civilizations, theories about fantastic voyages and colonization efforts, theories about cosmic events inspiring the mythic texts of the ancients, making them literally true, and theories about ancient astronauts. Unlike normal scientific paradigms, which ultimately must all fit together (a revision in the dating of Mesopotamian history must be shown to be compatible with what is known about Egyptian dating, for example), there seems to be no need for overall agreement among pseudoscientists themselves—was the Piri Reis map based on maps drawn with alien help, as von Däniken proposes,[1] or by an ancient advanced Antarctic civilization, as Graham Hancock suggests?[2] Did Venus erupt from Jupiter after an interplanetary collision and swing by the Earth to cause the catastrophes described in the Hebrew Bible, as Velikovsky argues,[3] or did the planet Marduk smash into the planet Tiamat to create the Earth and the Moon—Venus already existed—as Zecharia Sitchin tells us is recounted in the Babylonian creation epic *Enuma Elish?*[4]

Pseudoscientists also frequently fail to keep up with current research. Because the pseudoscientific paradigm is so rigid (see the Introduction), not only does it not tolerate the scientific paradigm it desires to replace but also tends to ignore developments that have taken place within that rival paradigm. There is a certain logic here, of course: Since the rival scientific paradigm has already been largely or completely rejected by the pseudoscientist, any future research done within it is regarded as false by definition and therefore can be ignored (unless, of course, some piece of data can be extracted and used to support the pseudoscientific paradigm in some way). The result is that the pseudoscientist tends to fight the same battles, using the same often outdated evidence, again and again. An excellent example of this are the ancient astronaut ideas of von Däniken from his 1970 book *Chariots of the Gods?* (first published in German in 1968). In 2010 the History Channel series *Ancient Aliens* rehashed many the same arguments, paying virtually no attention to newer evidence or more recent interpretations of this material.[5] Always remember: In mainstream scholarship, one is obliged to be familiar even with the views that one disagrees with.

This raises another question: how does the pseudoscientist arrive at their paradigm in the first place? Here we see the importance of *mythos,* the emotional sense of meaning. Paradigms of all sorts are often reached in flashes of insight, and so they are often accompanied by strong emotional power; who can forget the story of Archimedes leaping from his bathtub shouting "Eureka! Eureka!"? Science is designed to remove that emotion by testing the insight against evidence to see if it still holds up, and if it doesn't, it is dismissed. But in the pseudoscientific paradigm, the *emotional* impact of the insight is so strong for the person that they cannot disassociate it from the idea, and so retain it; in many ways it becomes little different from a religious experience. And like many *mythos* ideas, particularly those which claim exclusive truth, the pseudoscientific paradigm is then seen as a thing that must be defended against rivals. We will return to this below.

1.2

Here Silly uses large numbers to demonstrate that he has done some basic research, but he fails to distinguish between facts and speculation. The number of stars in the Milky Way galaxy and the number of galaxies cited here are estimates within the common ranges given by astronomers, and the discovery that extrasolar planets are common is correct, but the statement that earthlike planets are also common is still speculation. In fact, most extrasolar planets, like most of the planets within our own solar system, are not suitable places for life as we know it.

Silly also begins by estimating the number of planets in the universe and then jumps without warning to his estimate of the number of intelligent species in the Milky Way galaxy. This gives the false impression that he is being very conservative in calculating the likelihood of intelligent life. The estimate of the number of intelligent life forms in the galaxy is based upon a particular set of variables plugged into what is known as the Drake Equation. Developed by the astronomer Frank Drake in the early 1960s, this is a theoretical framework by which the number of sentient, technological life forms could be estimated:

$$N = N* \cdot f_p \cdot n_e \cdot f_l \cdot f_i \cdot f_c \cdot f_L$$

$N*$ represents the number of stars in the galaxy (or, if you like, any region of space), f_p the number of those stars that hold planets, n_e the number of those planets that can sustain life, f_l the number of those planets where life actually develops, f_i the number of those planets where intelligent life appears, f_c the number of those intelligent life forms that choose to communicate with other intelligent life forms, and f_L the period of time communicative civilizations actually communicate. Save for the first variable, there is little agreement about the values we should use. If you plug in the right numbers, Silly's figure of 100,000 is possible; change the values, and it can either rise or fall.[6] But Silly fails to either cite or explain the Drake Equation or how and by whom the figure of 100,000 was reached. Instead, were are simply told that these are numbers, and numbers don't lie. Silly also implies that all 100,000 other intelligent species in the galaxy are "close by," when in fact the Milky Way is 100-120,000 light years in diameter, meaning that even if his estimate is correct, most of those species will be very far away. Erich von Däniken makes similar use of the formula, stating, "According to this formula there are at any moment in our galaxy alone 50,000,000 different civilizations which are either trying to get in touch with us or waiting for a sign from other planets."[7] In fact, the equation does not assert this at all. It is simply a theoretical framework for estimating how many intelligent species the galaxy *may* hold, and it says nothing about whether or not extraterrestrials desire contact with Earth, or, if you're a *Star Trek* fan, want to date Captain Kirk. Once he has given the number of 50,000,000 as the single correct result of the equation, von Däniken then explains that the number may vary, and that 50,000,000 is the "admissible maximum value," and that the minimum value is 40.[8]

Note also Silly's statement "it is undeniable by even the simplest mind." This is a common feature of pseudoscience, qualifying an assertion with a statement that belittles any who might not agree. Von Däniken makes a similar implication when discussing the famous sarcophagus lid of king Pacal of Palenque:

> A genuinely unprejudiced look at this picture would make even the most die-hard skeptic stop and think.
> There sits a human being, with the upper part of his body bent forward like a racing motorcyclist; today any child would identify his vehicle as a rocket.[9]

Since no one wants to be seen as stupid, the reader is encouraged to accept von Däniken's assertion at face value. This is a subtle form of intellectual bullying, meant to sway a reader rather than illuminate a piece of data. In our parody, Silly's argument that his assertion has been reached "by the calculations of our most brilliant minds" hopes to reinforce that assertion, even as it gives no indication of who these brilliant minds actually *are* (and why are they all *men?*), how they supposedly arrived at these numbers, or where the reader can go to learn more. Instead, we are expected to simply accept the author's statement (again, real or parody) without question; a classic feature of pseudoscience. Even if the assertion were true this would be a dreadful violation of normal scientific protocol, which exists to minimize bias, not exaggerate it.

1.3

Here we see another common feature of pseudoscience: the romantic notion of an independent mind unfettered by the "experts" or by their evidence.[10] A scholarly conspiracy is implied, and the conservatism of careful scholars is turned into a liability. This has the effect of diminishing the scholarly community, and scholarship in general, allowing the pseudoscientist to become the real "expert" without needing any serious training, while the person who has studied the problem in careful detail loses legitimacy precisely because they have *not* jumped to hasty conclusions. The misunderstanding of the scientific method by Silly and other pseudoscientists is a perfect example of what is called the Dunning-Kruger Effect, a term from psychology that describes how individuals overestimate their own cognitive abilities when dealing with subjects where they have limited expertise. Since we all have areas where our expertise is limited (I can't do brain surgery, for example), this is something we should all be careful of.[11]

Note also the use of quotation marks. When the "experts" say something that Silly agrees with, there are no quotation marks; when they say something he doesn't, the quotation marks go up. This implies that scholars whose conclusions Silly can use in support of his argument know what they're doing, while those who might disagree with him do not. A similar and subtle tactic is the use of italics for emphasis: "Science has *spoken,* has *decided.*" This gives the impression that the scholarly community believes that it has reached the final possible conclusion about a subject, when in fact the scientific method is founded on the idea that no conclusion is ever final and that all must be supported by all available evidence.

As many pseudoscientists do, Silly also cites cases where the maverick turns out to be right, as with Einstein, Schliemann, and Galileo. He does not mention, however, the vast number of people who proposed ideas and theories that turned out to be wrong, and also implies that scholars never recognize when a new explanation for a phenomenon is an improvement and adopt it, as happened with all three of his examples. Instead, Silly shows a complete ignorance of the process of scientific evolution, by which scholarly communities recognize anomalies in their paradigms, propose new paradigms to explain these, and adopt new paradigms when they work better than the old.[12] Contrary to Silly's claim, science, and indeed scholarship in general, is in a constant state of evolution, and scholars can be and are frequently convinced of new theories, provided those theories pass rigorous tests with evidence.

Silly is not the first pseudoscientist to drop Einstein's name into his work. Graham Hancock refers to it in reference to the work of

Charles Hapgood, on whose speculations regarding ancient maps many of his own theories are based.[13] Given that Einstein was a physicist, it is not clear why his endorsement of a work on history and mapmaking should be considered authoritative. It's also worth noting that Einstein, and later Hapgood himself, expressed doubts as to whether the weight of ice at the poles could cause the continental crust to shift.[14]

Speaking of history, Silly also misrepresents that. Einstein fled Germany in 1933 to escape the rising anti-Semitism of the Nazis, not because of a hostile scientific establishment. He had received the Nobel Prize in 1921, so it is hard to imagine that by 1933 he could have felt persecuted by his peers. Schliemann did encounter resistance to his claims about Troy,[15] and so has been trumpeted as the archetypal example of the lone genius triumphing against the stodgy academic establishment. What is overlooked, however, is that Schliemann's views (or some of them, anyway; he was also frequently wrong, and in fact it was Dörpfeld who finally identified the actual layer of Troy) were corroborated by later archaeology, meaning that their discoveries actually did constitute a paradigm shift. Of course, this does nothing to prove that other unorthodox theories will do the same.

And again, Silly switches almost at random between praising science and condemning it, wanting the prestige of science when it supports him but attacking it when it does not (Nigel Davies noted this with von Däniken, who praises physicists and astronomers but condemns archaeologists.[16]). Further, the notion of the unfettered genius who turns out to be right has nothing to do with Silly's hypothesis, which at this point has not even been stated. Instead, he continues to set up the reader by making an emotional appeal and denigrating his critics in advance.

1.4

Again with the quotation marks around the word "experts." Silly's argument that follows is one of the most common in pseudoscience, found in everything from ancient alien belief to creationism to

Holocaust denial: *That negative evidence proves a positive assertion.*[17] The logic works like this: Since science is fundamentally based on the ability to admit when you don't know something and to admit when your previous thinking was incorrect, it lacks the ability to actually prove anything and therefore cannot be true. We don't know *exactly* how the Egyptians built their pyramids, for example, and so all theories about their construction must be wrong.

It is a short hop from this to the argument that if science or scholarship cannot prove something (negative evidence), then the theory of the pseudoscientist (a positive assertion) must therefore be correct. Another popular example is creationism: Since we do not know *all* the details about how biological evolution works, then the universe must have been created by God some 6000 years ago. Or aliens (or Atlanteans, or giant doughnuts) built the pyramids.

That this is logically inconsistent should be obvious. First, the fact that a person doesn't know something says nothing at all about the origin or cause of that thing. As I mentioned above, I know very little about brain surgery, yet brain surgery was still invented and is still performed. And second, the fact that one person (or even every person) doesn't know something doesn't mean that someone else's explanation about it is therefore correct. The two things are unrelated, and contrary to Silly's assertion, *my* lack of knowledge doesn't mean that *you* are right. Michael Shermer correctly calls this the "either-or" fallacy.[18]

Let's take this a little further. In sharp contrast to the claims of pseudoscientists (and the occasional overenthusiastic scientist), science does not claim to have found The Truth about anything. Rather, it uses evidence and logic to inductively propose explanations for phenomena, and evidence and logic to test those claims deductively. The explanations that emerge from this process are always tentative, subject to revision and even rejection by later scientists should they prove unable to explain the phenomena in question. In other words, every theory must stand or fall based on its merits, and an honest scientist must be able and willing to admit to the limits of their understanding. This admission is a sign of maturity, not the weakness of a particular scientific idea or the scientific method in general. The argument that your opponent's admission of their limits proves that a differing view is right, on the other hand, is a sure sign of pseudoscience and intellectual dishonesty.

1.5

Again we note the not-so-veiled insult directed against anyone who might disagree with the author's assertion: "the prejudices of small, primitive minds." The point that no evidence for extraterrestrial intelligence has yet been found is turned into a statement that "Silence means nothing." This is nonsensical and inaccurate, since astronomers have been looking for evidence of life on other worlds for decades.[19] Further, the analogy with the Apollo moon landings is misleading. The fact that human beings have been to the moon is not evidence of extraterrestrial intelligence; the two, while both interesting, are unrelated.

1.6

Here we see a classic feature of pseudoscience when it deals with the ancient world: the assumption that ancient human beings were "primitive" and therefore incapable of achieving anything impressive on their own. Silly is clearly following the argument of Zecharia Sitchin in *The 12th Planet,* though as is common with pseudoscientists, Silly does not cite his source.[20] The "primitive ancients" view is based partly on the assumption that the rapid technological growth that followed the emergence of complex states beginning in roughly 3000 BCE occurred in isolation and not, as argued by Jared Diamond in his book *Guns, Germs and Steel,* in response to evolutionary pressures such as population growth and the fortunate availability of domesticatable plants and animals in certain parts of the world.[21] Indeed, those pressures and assets are well attested in the archaeological record.[22] Instead of looking at current research on the subject, however, Silly argues that the emergence of complex societies could not "have been evolution, for evolution is random." This demonstrates a misunderstanding of the basic tenets of Darwinian theory, found also in creationism: that there is a particular direction or goal to the design of life forms, particularly human ones. What Darwin actually said was that through variations in organisms, those members of a species better suited to a particular environment will be more successful than those less so adapted, and will be more likely to pass favorable characteristics on to the next generation, resulting in movement toward adaptive success over time.[23] This is not a random process, nor is it a designed one, since environments and therefore what constitutes evolutionary success will change over time. The dinosaurs, for example, were very successful for many millions of years, until something changed (likely a series of catastrophic volcanic eruptions or a comet/asteroid impact, or both) and most of them went extinct (it turns out that birds are the descendents of the few who survived).[24]

A far more reasonable scenario than extraterrestrial interference is that the long period of slow technological growth for human beings is not surprising, since rapid technological growth was not necessary for the Darwinian success of early *Homo Sapiens*. It was only the unique set of circumstances that occurred after the last Ice Age that changed the environmental pressures on human beings and pushed them towards increasing technological and political/social complexity. The statement that something happened with the emergence of the Neolithic that spurred the development of complex states is correct, but the statement "We must have had help," is an assumption and is presented here without any evidence to support it.

1.7

The use of analogy is a common feature of pseudoscience, and it does have some value in legitimate scholarly investigation and presentation. However, analogy also carries risks, since the issues being compared need to be similar enough in the right way to avoid the old apples-and-oranges problem. In this case Silly begins with the assumption that extraterrestrial contact occurred in the past, even though he has not demonstrated that this assumption is valid.

Further, he assumes that von Däniken's assertion is correct, even though we have no case studies of extraterrestrial contact to use as evidence. He does not tell us where von Däniken made his statement (I found one such reference on page 10 of *Chariots of the Gods?*, though there are doubtless many others), but at least Silly gives him credit, probably as an appeal to authority in the community of ancient astronaut believers, where von Däniken is a major figure. The statement by Arthur C. Clarke is accurate, but hardly constitutes evidence, since it is also true that magic is used in a wide variety of contexts in human society having nothing to do with technology.[25]

So we have reached the end of the first chapter, and no thesis has been given. Presumably the last paragraph will lead us to the author's point.

2.1

There are several problems with this section. First, Silly quotes the *National Geographic* reference as an "archaeological report," when in fact it is merely a photograph with a caption.[26] Second, the caption states only that the origin, age and purpose of the stone circle are unknown, and it says nothing about the supposed archaeological speculations given here. Since Silly does not cite any actual archaeologists for these theories, we have no way of knowing if they have actually been proposed, or if he is merely inserting interpretations for other sites, or inventing them from scratch. What we see here is in fact a common technique used to take control of arguments: you state that your opponent has said something that they in fact have not. Erich von Däniken makes a similar statement in *Chariots of the Gods?*, where, in reference to the Nazca lines in Peru, he tells us: "The archaeologists say that they are Inca roads."[27] As Ronald Story points out, however, the first archaeological report on the lines explicitly stated they were *not* Inca roads, and that has remained the mainstream position ever since.[28] Silly's implication that archaeologists merely come up with theories willy-nilly and without careful research is similar to a charge made by von Däniken, worth quoting in its entirety:

Scholars make things very easy for themselves. They stick a couple of old potsherds together, search for one or two adjacent cultures, stick a label on the restored find and—hey, presto!—once again everything fits splendidly into the approved pattern of thought. This method is obviously much simpler than chancing the idea that an embarrassing technical skill might have existed or the thought of space travelers in the distant past. That would be complicating matters unnecessarily.[29]

The simple fact that archaeology doesn't actually work that way is ignored, and yet von Däniken expects the archaeological community to take him seriously. What is particularly ironic is that it is von Däniken and other ancient astronaut believers, not professional archaeologists, who argue that ancient peoples lacked technical skills.

Silly's second paragraph has a careless error: the nomads of the Sahara are primarily Taureg, not Bedouin (Bedouin live in the Middle East, not Africa. Had Silly read the actual *National Geographic* article he would have seen this.). Finally, the reference to Potemkin's village implies that archaeologists are intentionally misleading the public, a serious charge that Silly fails to substantiate.

2.2

Silly ignores virtually all the research ever done on Stonehenge, relying on the impressive nature of the site to convince the reader that its construction was beyond the capability of the people of prehistoric England. This is a classic argument of pseudoscientists in dealing with large-scale monuments: because they are impressive feats of engineering, they are beyond the ability of "primitive" peoples to create. Yet numerous experimental efforts by researchers have shown that even relatively small numbers of people with ropes can move multi-ton stones long distances.[30] Silly states that all of the stones weigh 50 tons, when in fact only the heaviest do. For some reason he also argues that modern technology has difficulty moving even a one-ton block of stone, when in fact we do it all the time.

It is necessary to maintain the argument about all prehistoric and ancient peoples being "primitive," of course, since if the pseudoscientist acknowledged that ancient peoples were inventive problem solvers (as stone-age peoples have repeatedly been shown to be in modern ethnographic studies[31]), then he must also entertain the possibility that such people could create the impressive monuments attributed to them, and this weakens the original pseudoscientific view.

2.3

Here we see another assumption common to the pseudoscience of ancient astronaut belief, that any attention paid to the sky or heavens implies extraterrestrial visitation. Yet a number of cultures and individuals for whom we have good records merely observed the sky out of interest, and their records say nothing about extraterrestrials—Europe comes to mind—which calls the implication into question. That extraterrestrials would come from the sky does not mean that all interest in the sky involves extraterrestrials; this is an assumption only.

The quote from *National Geographic* is incomplete and taken out of context. This is another common feature of pseudoscience, and combined with the frequent lack of good citations (or sometimes any citations at all), is a way to make a source say what you want it to say while at the same time making it difficult to uncover your deception. For the record, here is the entire sentence:

> Skeletal remains indicate that despite physically demanding lives, the people of Neolithic Britain were more lightly built than us. Their relative lack of dental decay suggests a diet low in carbohydrates, and although life expectancies are difficult to calculate, people seem, overall, to have enjoyed good health.[32]

This is telling. Silly argues that the ancient Britons were hungry for more calories, implying a need for an improved diet, but the quote he cites to support this actually says the opposite, that the low calorie diet resulted in fewer dental problems and general good health.

A nice summary of the subject of Stonehenge can be found in Peter Lancaster Brown's 1976 book *Megaliths, Myths and Men*, which discusses the astronomical alignment theories in some detail. A good discussion of the monument in relation to various religious and pseudoscientific theories appears in Kenneth Feder's 2002 book *Frauds, Myths and Mysteries: Science and Pseudoscience in Archaeology*.

Flashes of insight are well-attested in scientific circles, with those of Archimedes and Newton being among the most famous, though the story of the apple actually hitting Newton on the head seems to be apocryphal. A particularly amusing and more recent anecdote is that of George Smith, who while reading newly discovered cuneiform tablets in the British Museum in 1872 found the flood story from the Epic of Gilgamesh and realized its similarity to the Noah story in the Bible.[33]

The point, of course, is not that one has sudden insights, but what one does with them. Any insight must be tested against actual evidence using a methodology designed to eliminate bias, and only when current explanations prove inadequate should new ones be explored. Once again, this sort of approach is usually lacking in pseudoscience, where the clever idea is more important than the evidence itself.

3.1

It is difficult to make sense of this section, but it does reveal an important point about pseudoscience that we will see again and again: While it claims to make use of objective, rational thought, much of it is shrouded in mysticism. The value of mysticism in scientific inquiry is limited, since the mystical makes use of elements of the human mind that can neither be tested nor always shared. Mystical experiences tend to be unique and/or uncontrolled occurrences, and if you can't repeat them or put them into controlled conditions, you can't evaluate them scientifically. The advantage of mysticism for the pseudoscientist is significant, however, since mystical experiences carry powerful meaning to those who have them, and they can also create powerful emotional bonds between people. This means that belief in a pseudoscientific mystical system can bring membership in an exclusive group who have shared or want to share such experiences. The powerful appeal of mysticism likely reflects our right brain, which is less detail oriented and seems to seek meaning. So long as mysticism is not used as scientific evidence, and so long as it is understood as part of our deeper and often irrational minds, it is not terribly harmful and probably is even beneficial. Certainly we are left to wonder why the irrational and the mystical side of our minds would have evolved if they were not adaptive in some way, and cases where the irrational parts of our brains are damaged or shut down show serious impairment, particularly in the ability to actually act on a decision. In its own way this is just as harmful as when the rational parts of the brain fail, perhaps even more so.[34] So Silly's case may be, well, silly, but that doesn't mean that other mystical experiences are.

The material presented by Silly resembles the writings of Helena Blavatsky, who founded the movement of Theosophy and whose most famous work, *The Secret Doctrine,* purports to be an Atlantean book called the *Book of Dzyan,* written in a secret language called "Senzar" (I swear I'm not making this up). Here's an abridged example:

> 1. The Eternal Parent, wrapped in her Ever-invisible Robes, and slumbered once again for Seven Eternities.
> 2. Time was not, for it lay asleep in the Infinite Bosom of Duration.
> 3. Universal Mind was not, for there were no Ah-hi to contain it.
> 4. The Seven Ways of Bliss were not…The great Causes of Misery were not…for there was no one to produce and get ensnared by them.[35]

What this actually means, of course, is really anybody's guess, and again, this is the hazard of using mystical language and experience in scientific inquiry, since mystical experiences tend to be highly personal and subjective. The parts from the *Book of Dzyan* are accompanied by Blavatsky's own commentary, much of which attacks both science and religion.

Returning to Silly's work, what is most striking about this section is how the opening paragraph makes claims for seeking evidence with scientific rigor, and yet Silly then quickly turns to "Oriental" mysticism to begin his investigations. The concept of Orientalism, most famously raised by Edward Said, may be broadly described as the belief in Western culture that the "Orient" (here defined as the Middle East and India) is a dark, mysterious place, sensual and decadent, but also possessing ancient wisdom and mysteries.[36] It is a fantasy view that rarely has any relation to the reality of life there, but it does help sell books; Americans would be much more likely to buy *Love Secrets of the Orient* than *Love Secrets of Topeka, Kansas*. This appeal is found in numerous pseudoscientific works; probably the closest parallel to Silly is James Churchward, who in his book *The Lost Continent of Mu* describes meeting a wise man in India who taught him the secret language of the lost civilization of Mu and then introduced him to a set of ancient tablets that told of that lost civilization (a sunken continent in the Pacific).[37]

3.2

Again Silly cites part of a text out of context, without providing his source. The text in question here is Svetasvatara, from the Upanishads, one of the sacred texts of Hinduism. The wheel analogy is used to describe the Hindu cosmological belief in cyclical time and reincarnation. The full passage reads as follows:

> This vast universe is a wheel. Upon it are all creatures that are subject to birth, death, and rebirth. Round and round it turns, and never stops. It is the wheel of Brahman. As long as the individual self thinks it is separate from Brahman, it revolves

upon the wheel in bondage to the laws of birth, death, and rebirth. But when through the grace of Brahman it realizes its identity with him, it revolves upon the wheel no longer. It achieves immortality.[38]

The primary logical problem with Silly's approach here is that it is deductive rather than inductive. Since he is looking for circular things and since he has already concluded that the presence of circular things indicates the influence or presence of his intergalactic doughnuts, he becomes trapped in circular reasoning (pardon the pun). Rather than follow the evidence, he leads the evidence by only seeking things he feels will support his argument. In this case he refuses to believe in the possibility of coincidence, even though a shape as basic as the circle can appear for many reasons.

Virtually every pseudoscientist falls into this trap. So if von Däniken reads a part of the Epic of Gilgamesh that sounds to him like it depicts the G-forces felt by an astronaut when a rocket takes off, that must be read both literally and only in that way. Other explanations, like the sensation of the heavy body of a god or a heavy person sitting on you, are not even discussed. If gods dwell in the skies, this can only mean they are ancient astronauts, not that the sky is part of the world visible from ancient Mesopotamia and that gods are everywhere (von Däniken seems far less interested in other gods who dwell underground). He argues later that "When the same tablet tells us that a door spoke like a living person, we unhesitatingly identify this strange phenomenon as a loudspeaker."[39] Why? Couldn't the god simply be speaking through a closed door? But alternate explanations do not interest von Däniken; there is his interpretation, and *only* his.

3.3

In these paragraphs we see a number of the dangers of relying too heavily on deductive reasoning when evaluating real-world evidence. Silly takes the existence of Atlantis as described in Plato at face value, without evaluating its context. (The story appears in

Timaios and *Kritias,* two dialogues where Plato is attempting to make a philosophical point.[40]) Since Plato's primary methodology was itself deductive, he had little interest in gathering observational evidence to support his positions. From his text we see no evidence that he ever attempted to verify the story, and it is entirely possible he made the whole thing up. As L. Sprague de Camp has noted, there is no mention of any sunken island in the Atlantic Ocean by any writer before Plato, which does little to support Silly's argument.[41] It is not unreasonable to say that a landmass that size and a civilization that important should have appeared in some other record. As Marcello Truzzi put it in a phrase made famous by Carl Sagan, "Extraordinary claims require extraordinary evidence." Pseudoscientists are very good at the first part, but not so much the second.

Much of what Silly says about Atlantis is clearly derived from the work of Ignatius Donnelly, who argued that, just as Plato said, the continent of Atlantis had been located in the middle of the Atlantic Ocean and that it was the origin of civilized cultures in both the Old and New Worlds, as well as the Garden of Eden, the Gardens of the Hesperides, the Elysian Fields, the Gardens of Alcinous, the Mesomphalos, Olympus, and Asgard. It was also the origin point of many of the deities of the ancient world, and it "perished in a terrible convulsion of nature," sinking into the sea, an event which in turn spawned all the ancient flood stories.[42] We should not underestimate the importance of this work or its influence on later pseudoscientists or the popular imagination. Even today, a new "discovery" of Atlantis seems to be announced every few years, and the lost continent has been placed on virtually every piece of real estate on Earth.[43]

Silly also makes a number of erroneous statements in his discussion. Egypt is one of the oldest civilizations known to us, but not the single oldest. This is a common error in many of the more mystical pseudoscientific ideas of the past century, and one of these was probably the source for his claim. More serious is Silly's rejection of modern evidence regarding the Atlantic seafloor, where in one paragraph he states that scientists have mapped the area, and in the next denies that they are able to do so. He follows this with the common pseudoscientific view that when the experts disagree with him they are not only wrong but disconnected from reality in their "Ivory Towers." Finally, he assumes that what "Asia" meant to Plato is the same thing it means to us: the largest continent on Earth. From this Silly constructs an argument that both North and South America are part of the former Atlantis continent, allowing him to claim that Native American cultures are descended from Atlantis. His further claim that we know "next to nothing" about the Maya and Inca is simply not true; both have been extensively studied.[44] Silly is also not the first person to claim that the Americas are the remains of Atlantis. Ronald Fritze tells us that the idea was common among Atlantis believers for more than two centuries, beginning in 1530 with Girolamo Fracastoro and lasting until the 19th century.[45] Reading multiple pseudoscientists, we find that there are very few new theories among them, but lots of old ones recycled.

As we saw above with Orientalism, the Maya and the Egyptians are appealing to pseudoscientists precisely because so much about them seems strange and different, and because only a small fraction of their culture has survived. This allows the imagination to fill in the gaps with good stories. The more they are studied, of course, the more like the rest of humanity they prove to be. To my mind, this makes them more interesting, not less, but to each their own.

3.4

It's tempting to say that the less said about the ridiculousness of this argument, the better, but it actually does illustrate some important features of pseudoscience. Let's review: The Greek omicron (**O** or **o**) is derived from the Phoenician letter *ayin*, familiar from Hebrew or Aramaic. And while *ayin* does appear as a circle in Phoenician and paleo-Hebrew, in later Hebrew and Aramaic it appears as ע, and in the older alphabetic script of Ugaritic cuneiform as ᛐ. Zecharia Sitchin makes a similar argument to Silly's using cuneiform signs in *The 12th Planet*, where he argues that the DIN sign connects with the GIR sign "just as the lunar module was docked with the Apollo 11 spaceship!"[46] Fritze has argued that "Sitchin's assignment of meanings to ancient words is tendentious and frequently strained",[47] which is a kind way of noting that Sitchin seems to believe that he alone can accurately read Sumerian. And while Sumerian is a very difficult language that is not fully understood,[48] the reading of the word for "god," DINGIR, *is* well understood, and it is *not* written as Sitchin implies. Here are the signs as Sitchin sees them:[49]

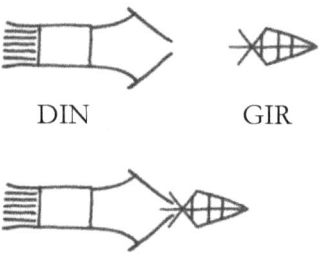

DIN   GIR

Because Sitchin fails to use the standard Assyriological references for these two signs, it is very difficult for a non-specialist to check him. Consulting these sources, we find that his DIN is actually SIMUG, occasionally representing DIM (as DIM_6), but that it is *not* listed with the phonetic value DIN. His GIR is listed as GIR_2.[50] Proper citation is a common problem with pseudosciences, and it is a rather serious one, since citations are intended not merely to give credit for the ideas of others, but also to allow the reader to check the accuracy of the writer, as I have done here. Of course, since pseudoscience is primarily concerned with defending the paradigm (there's that issue again), it is actually important for the pseudoscientist to make it *harder*, not easier, to check their arguments for accuracy (*Chariots of the Gods?* uses virtually no direct citations at all). Fritze notes that in the case of Sitchin, those who do check his sources frequently find that "he quotes out of context or truncates his quotes in a way that distorts evidence in order to prove his contentions."[51]

My analysis of this instance bears this out. Sitchin neglects to mention here that the Sumerians almost always wrote "god" with one sign, DINGIR, not with two, and that the DINGIR sign does not represent a spaceship but a star, something that would have been readily visible in the night skies of southern Mesopotamia. To argue that the combination DIN+GIR is a pictorial representation of a rocket or that the name Poseidon is related to doughnuts based on sign shapes really only shows the silly places that an overreliance on coincidence can take you. Kenneth Feder has commented on the phenomenon of the inkblot test, in which shapes are assigned meaning based on the beliefs of the viewer, not on their actual form, which may be entirely random.[52] When you approach things deductively, you will already have an idea of what you are looking for in your evidence, and one of the risks is that you will naturally start to see what you want to be there. But desire is not evidence, and hunches such as Sitchin's and Silly's are only useful if they are then verified *with* evidence, not imaginary flights of fancy.

3.5

The idea that the Aegean island of Thera is the inspiration for Plato's Atlantis story has been around for some time, since archaeologists discovered the remains of Minoan settlements there. But note how Silly focuses on what is similar and not what is different, and how he uses language to avoid facts. The evidence is that most of the people of Santorini got off the island before the volcano there exploded, as few skeletal remains were found by archaeologists (the Minoans were clearly more cautious than the Romans at Pompeii).[53]

Further, Silly misuses geology in describing volcanoes. There are basically two types: one, found over geological "hot spots," tends to produce mountains by a slow accretion of lava, and Hawaii is a good example of this. But the other sort, often found where one geological plate is subsumed beneath another, tends to have periodic violent explosions such as that of Mount St. Helens in 1980 or Krakatoa in 1883. What Silly is doing here is an obvious and egregious example of using facts only when they support you, and ignoring them when they don't, something no reputable scholar would ever intentionally do.

The double standard appears again in the next paragraphs. Note that Silly calls the people of ancient Thera "ancient and mysterious," and yet he argues that Plato knew about them, speaking with some authority. So things are mysterious when they suit one part of the pseudoscientist's narrative, and yet are crystal clear when they support another.

This sort of double standard appears in other pseudoscientific claims. A really good example can be found in the argument by Robert Bauval and Graham Hancock that a global catastrophe caused a shift in the Earth's crust in about 10,000 BCE, which moved the locations of the north and south poles. But as James and Thorpe point out, this argument for a shift in the Earth's crust is based on the locations of the stars of the zodiac going back to about 10,500 BCE, and assumes that Egypt and other early civilizations were at about the same latitude they are now, which would not have been the case if the Earth's crust underwent the massive shift Bauval and Hancock claim. "One cannot, as Hancock tries to," James and Thorpe remind us, "have it both ways."[54]

Note how again Silly concludes his discussion with loaded questions and the same comment about scholars being trapped in their "Ivory Towers," both of which are intended to lead the reader into agreement rather than an evaluation of the evidence.

3.6

Everything in these last two paragraphs rests on undemonstrated assertions: the existence of Atlantis, the existence of the circular city of Atlantis, and the ability of pastries to sink continents in such a way that leaves no trace in the geologic record. The assertions do, however, demonstrate another feature of pseudoscience, which is to conclude that the assumption of the previous statement has been proven and can therefore be used without question in future claims, resulting in an increasingly improbable set of hypotheses all tied together to give the illusion of a coherent theory. Both Erich von Däniken and Graham Hancock use this technique extensively.[55] Doing this also buries the potential critic under a pile of information, presented quickly and in such a way that he or she has no time to evaluate individual claims before being pushed on to the next assertion.

The use of television makes this trick even easier to pull off, since unlike a book you cannot easily flip back to previous pages; DVDs are far less user-friendly this way than a written text. The History Channel series *Ancient Aliens* employs it with considerable sophistication, leaping rapidly from one topic to another and considering none of them in any sort of depth. This is odd considering how the first season alone is nearly eight hours long, more than enough time to consider some of its claims in detail.

3.7

If Silly is referring to the Cueva de los Toyos under a different name, his description is completely inaccurate, and calls into question whether the visit described here ever actually took place. The cave is located near the bottom of a valley, not on a hilltop. What is also noteworthy is that Silly has clearly taken his own visit from the account of Erich von Däniken, who made similar claims about metal tablets and golden artifacts there in his book *The Gold of the Gods*, a position he later recanted, claiming that "In German we say a writer, if he is not writing pure science, is allowed to use some *dramaturgisch Effekte* – some theatrical effects. And that's what I have done."[56] Taking this sort of liberty with the evidence is odd, however, if von Däniken wants his ideas to be taken seriously by the scholarly community, since that community tends to frown on researchers inventing data.

However, even if von Däniken's claims are fictional, their lack of mention by Silly is improper, since by ignoring von Däniken Silly implies that he is the first researcher to enter the Cueva de los Toyos (or Pendejos) in search of alien artifacts, which it is safe to say he is not. Of course, considering that von Däniken himself has repeatedly been guilty of similar omissions of the work of previous ancient astronaut advocates such as Louis Pauwels, Jacques Bergier, and Robert Charroux, perhaps Silly is simply emulating the ancient astronaut movement's acknowledged master.[57]

Following von Däniken's purported visit, the caves were investigated by Stan Hall and a large team, who found no evidence to support most of von Däniken's claims, though Hall, himself a pseudoscientist, still claims that the library is there somewhere.[58] This expedition also made no mention of a "Cueva de los Pendejos," which, to the best of my knowledge, is only mentioned by Silly. There is a Pendejo Cave in New Mexico, which has produced Clovis period artifacts, and perhaps Silly is merely confusing the two. The guide "Juan Juarez" is also not mentioned in other accounts of the caves, though a "Juan Moricz" is prominent in Hall's account, where he is said to have been highly critical of von Däniken.[59] A similar critical attitude towards von Däniken is recounted in an interview granted by Moricz to *Der Spiegel* and quoted at length by Ronald Story, who also notes other critics of von Däniken and his alleged source Father Crespi.[60]

3.8

The reference to the Senzar language and the *Book of Dzyan* confirms the influence of Helena Blavatsky on Silly's thinking.[61] Here, however, we see another feature common to pseudoscience: evidence is claimed, but not produced, often in the form of ancient texts that are conveniently unavailable for study by other researchers. Naturally, the pseudoscientist can also read the unknown language of the alleged mystery texts, which invariably support the conclusions the pseudoscientist wants to reach. Other examples of such unavailable mystery texts are numerous, including James Churchward for the lost Pacific continent of Mu, and Joseph Smith, whose *Book of Mormon* is the second holy text (following the Christian Bible) of the Mormon religion.[62]

4.1

This fanciful account of an encounter with spacefaring dough-nuts is paralleled (though without the pastries) in *Chariots of the Gods?;* a similar and fictional prediction of the reaction of scholars follows.[63] It may also interest the reader to know that scholars live very much like everybody else, with things like houses and groceries and jobs. Some of us even enjoy reading and writing science fiction.

4.2

The statement that the wheel is critical to civilization is incorrect. New World civilizations achieved complex society in all its forms without using the wheel for much more than toys. The wheel is helpful, to be sure, but not essential. And the fact that the wheel is round with a hole in the center has nothing to do with pastries; this is simply the way a wheel must be shaped if it is to work. Further, how does Silly get from the usefulness of the wheel to the idea of extraterrestrial doughnuts? These sorts of leaps of logic are central to most pseudoscientific arguments, implying that two things are connected when there is no demonstrable reason they should be.

And again, how does Silly know that our ancestors were "igno-rant and savage"? This is a basic feature of ancient astronaut belief, and is repeated so often that believers have come to accept it as a given, from which all argument must follow. The truth is that there is no evidence whatsoever that ancient people were any more or less intelligent than modern ones, and the two world wars of the 20th century should put to rest any notion that we are less savage than we used to be.[64]

4.3

Ezekiel's vision is one of the most popular subjects for ancient astronaut believers, even to the point that one man actually patented the design of the wheels of the "spaceship,"[65] and so it would be surprising if Silly left it out of his own theory (von Däniken goes over the text briefly and without context in *Chariots of the Gods?*[66]). This does raise an important point: the idea common in pseudoscience that ancient texts should be read in a literal way, much as we might read the newspaper, and that the ancients therefore saw the world exactly the same way we do. This seems at odds with the equally common pseudoscientific belief that the ancients were stupid, unless you want to read these theories as cynical commentaries on the human condition.

These views ignore historical context. Ezekiel was one of the main prophets of the Babylonian Exile of the Israelites (c. 598-538 BCE), and his book is a carefully constructed argument that first condemns the Israelites for abandoning their god Yahweh and then brings them back to him, promising hope and restoration. Any reading of the text that does not take this into account must be considered flawed. Further, the pseudoscientists who claim that Ezekiel's vision is of a spacecraft make no mention of the wider context of the visionary imagery found throughout the Israelite prophetic tradition. How does Ezekiel's vision fit into that context? Why look only at Ezekiel's chariot, and not at the story of the dry bones being brought to life in chapter 37? Biblical scholars look at the whole text, but pseudoscientists, like religious fundamentalists, pick and choose what suits them for the moment.

4.4

Silly's use of the Roman Coliseum as evidence for extraterrestrial visitation is, to the best of my knowledge, unique in pseudoscience (though given the large corpus, I may have missed some). In this way, however, he is actually being less ethnocentric than other ancient astronaut believers who slant their evidence toward cultures from outside of Europe, as pointed out by Feder. Like Silly, Feder notes that the Coliseum is an architectural wonder just as the Pyramids are, though this may be the only point of agreement between the two. Feder also points out that the Parthenon in Athens and the Minoan temple at Knossos are architectural wonders that are not associated with extraterrestrials,[67] to which I add the observation that Knossos has no readable texts that document its construction, which you would think would invite such fantastic speculation.

Another point is made here that does appear in the work of von Däniken, who notes that the Nazca lines in Peru include images that can only be seen from the air. Because of this, von Däniken argues, they must have been created to signal beings that had access to flight.[68] Why sky gods who the designs might have been made for had to be extraterrestrials is not explained, since all that would actually be needed was the *belief* that gods existed in the sky, and human beings are perfectly capable of many beliefs without witnessing actual examples of them (extraterrestrial doughnuts come to mind). For this reason if no other (and there are many others), the shapes at Nazca cannot be considered evidence of extraterrestrial visitation.

Of course, the fact that the precise purpose of the Nazca lines is still a mystery leads pseudoscientists like von Däniken to run wild with speculation.[69] This raises again the unwillingness of pseudoscience to tolerate uncertainty even as it happily embraces mystery. Limitations do not sit well with our culture, in which we are told regularly that "the customer is always right" and that you can "get something for nothing," and it is a characteristic of pseudoscience to try and fulfill these desires. As a result, pseudoscientists imply that their theories can answer all questions related to their claims, and that this ability makes them superior to the paradigms used by scholars, who proceed much more cautiously before making conclusions and who ideally will readily admit what they do not know. David Childress, author of *Technology of the Gods,* has seized on this to attack traditional archaeologists (I have placed sections of the text in italics where the speaker's tone seems to indicate emphasis):

> One of the good things about ancient astronaut theory is that it makes us think. It brings out the "What about this? What about that?" We need explanations for all these things. And I think that one of the main reasons why mainstream archaeologists get a real knee-jerk reaction to the ancient astronaut theory is that it's very disturbing to them, because they *don't* have all the answers.[70]

Of course, there is a very real difference between being disturbed by not having answers and accepting proposals based on faulty evidence and methodology. The former encourages the scientist to do more research, while the latter encourages the pseudoscientist to accept things based on faith in a paradigm that claims to be inerrant. Again, the flawed logic being proposed by these sorts of arguments is that if a scientist does not know the answer to a question, the pseudoscientist's argument must therefore be correct, even if there is no positive evidence to back it up. I repeat for emphasis: You cannot prove a positive assertion (aliens built or inspired the Nazca lines) from negative evidence (we don't know the exact purpose of the Nazca lines). You need positive evidence, which pseudoscientists rarely if ever provide.

4.5

The Piri Reis map has engendered speculation by numerous pseudoscientists, from Charles Hapgood to Erich von Däniken to Graham Hancock, who argue that it accurately represents the coast of Antarctica without ice.[71] Others disagree, arguing that the map is not exact,[72] and in fact a close look at it shows several striking differences with actual coastlines, including:

1. Cuba is completely inaccurate.
2. The Strait of Magellan is absent and the southernmost part of South America curves eastward across what is in fact ocean.

Silly is in fact more accurate in his discussion of the map than von Däniken, since the doughnut theorist at least acknowledges that Antarctica (or the coastline claimed to be Antarctica) is too far north on the map. Silly borrows from Hapgood and Hancock the idea that Antarctica shifted rapidly southward, though he disagrees with them on when this happened.

This all raises the question of catastrophism versus uniformitarianism. The general view in the physical sciences is that most change in the natural world occurs slowly over great periods of time. In other words, the world of the past worked just as it does today, and it is hard to see the big changes like continental drift and even biological evolution. Catastrophism, on the other hand, argues that change occurs suddenly, with large-scale natural events that quickly alter the natural world. This idea was at the root of the theories of Immanuel Velikovsky, who argued that the events depicted in the Hebrew Bible and other ancient legends were literally true observations of such catastrophes. The theories of the sinking of Atlantis and other lost continents work from much the same premise.

The problem with the dichotomy between the two positions is that they are often presented as an either-or argument, rather than the possibility that both might be true. In fact, the evidence is that both are. Change does take place slowly, often too slow for humans to see, but it also happens quickly, overnight. The problem with catastrophism is not that it is wrong but rather that its adherents usually fail to consider the ramifications of the events they claim are real. As noted above, we do have good evidence of a worldwide catastrophe in the K-T event of about 65 million years ago, when an asteroid or comet between 10 to 15 kilometers wide impacted the Earth in the area that is now the northern Yucatan Peninsula, releasing a blast estimated at 100 million megatons and, according to computer models, causing earthquakes of 12 or 13 on the Richter Scale, a tsunami 100 meters high worldwide, and setting fire to virtually the entire surface of the Earth (keep this in mind the next time you think *you're* having a bad day).[73] But by the standards of Velikovsky, Hancock and Silly, this event was minor. The K-T asteroid/comet impact did not shift a single continent and did not stop the Earth from rotating. The kind of energy needed to do those sorts of things, not to mention to sink an entire continent overnight, is orders of magnitudes greater, and it is hard to imagine anything or anyone surviving that level of destruction to put it into an ancient myth. In other words, continental drift is slow, and that's a good thing. If it wasn't slow, we'd probably all be dead.

4.6

The problems with Silly's hypothesis here could fill a book. His dating is impossible; if the sources of the Piri Reis map are to be dated to 300 BCE, then how can this postdate Akhenaten by a full millennium, when we are told that the Egyptians *rejected* the ancient doughnuts? The idea that Akenaten was either an alien or a human-alien hybrid has also been raised in the History Channel series *Ancient Aliens,* because in his iconography, he does look rather odd, with wide hips and an elongated face, and some scholars have speculated that he may have suffered from a medical condition such as Froelich's Syndrome or Marfan's Syndrome.[74] More recently, however, the discovery of what seems to be Akhenaten's mummy indicates that he didn't look like his images at all, meaning that the peculiar representation of this pharaoh was an iconographic convention and not a reflection of actual physical deformities.[75] And we should not be surprised that the Aten is depicted as being round, since it was, after all, the physical disc of the sun in Egyptian religious belief. In fact, the only thing even remotely plausible about Silly's statements about Akhenaten and the Aten is the association of doughnuts with big hips, though not in the way he probably intends.

As to the "doughnuts" on the Piri Reis map, these are compass roses—features used on maps since the Middle Ages to show cardinal directions—and the lines that emanate from them are portolan markings that represent compass headings taken from a particular point.[76] But like everything else in the world that is round, Silly sees them as doughnuts. This is the "inkblot effect" described by Feder run amuck. Silly is clearly seeing what he wants to see, and since round images and objects are common, he will therefore see his doughnuts everywhere (I expect this may be more acute when he is hungry).[77] He then constructs hypotheses out of his imagination, ignoring any evidence that might indicate a different interpretation.

Finally, as to just what a "psychotronomatic ray" is, your guess is as good as mine.

4.7

The Maya have long been a staple of pseudoscience, in part because to the casual observer, they do present a mystery. Here Silly repeats arguments that are at the core of ancient astronaut belief, particularly that the descendants of the Maya (who, by implication, are primitive and foolish) could not have possibly been responsible for the brilliant things their civilization achieved.

The reason for this has nothing to do with the actual Maya, however, but rather the strong influence of emotion. So again, Mystery is a powerful thing, as any author of fiction can tell you; to give away the whole plot too early spoils the fun. As long as the Maya, or the Egyptians, or any ancient people can remain mysterious, we will be attracted to them, and the more scholars fill in gaps in our understanding, the less we can invent exciting stories to explain things. Sadly for pseudoscientists, we find that the actual Maya are like the actual Egyptians or Mesopotamians or peoples of Easter Island: they are human and not so different from us. This reality is unpalatable to the pseudoscientist, especially since understanding the past in such a sober way requires more training than they generally want to invest in, and the results tend to sell fewer books.

It is odd to scholars that pseudoscientists argue that what scholars do is dull, since the thrill of discovery is just as strong in actual science as it is in fiction. The Maya *are* fascinating; all humans are, and yes, they are still mysterious and will almost certainly remain so. Why you need aliens or Atlanteans or divine action to make the world interesting is perhaps the greatest mystery of all.[78]

Silly does inadvertently make an interesting point, however, which may explain, at least in part, why pseudoscientific theories are so popular. The American educational system, particularly in the arts, humanities and other fields that teach critical thinking, has been failing badly for decades. History textbooks often teach as fact things that are manifestly not correct, and students emerge both bored with history and believing the mistakes and distortions their textbooks tell them. The fact is that Columbus did not discover the Americas (he obviously could not have if there were people already

there, and we know that even among the Europeans the Norse pre-
ceded him by about five centuries). He was also not the first European to meet the Maya; that was Juan Ponce de León in 1513.[79]

The interstellar "Peace Corps" view has been articulated by
Kenneth Feder in reference to the theories of von Däniken.[80] Von
Däniken may also be Silly's source for the legend that Teotihuacan
dates to a time before human beings were created, since he states:
"The oldest text about Teotihuacán tells us that the gods assembled
here and took council about man, even before *homo sapiens* existed!"[81] The archaeological evidence from Teotihuacan, on the other
hand, indicates that the pyramids of the Sun and Moon were built in
the period between 100-500 CE, at which time the city was one of
the largest urban centers in the world. This raises problems of chro-
nology in relating these structures to the pyramids in Egypt, which
were built more than two thousand years earlier. The argument that
Mesoamerican pyramids are similar to Egyptian ones also does not
hold up, since the Mesoamerican examples have temples on top of
them and the Egyptian examples not only do not, but cannot, since
they end in a point.

The statement that Teotihuacan forms a map of the Milky Way
is completely unsubstantiated.

4.8

Here Silly has managed to integrate one of the earliest pseudoscientific claims about the Maya, that of Augustus le Plongeon
(1825-1908), with the later ancient astronaut ideas of Erich von
Däniken. Le Plongeon, in addition to doing valuable early photog-
raphy of several Mayan sites such as Chichen Itza and Uxmal,
claimed to be able to read the Mayan hieroglyphs. His "reading"
produced a fanciful tale of one Queen Moo, her two brothers, and a
trip to Egypt via Atlantis. Clearly le Plongeon has been consulted by
Silly here.[82]

The sarcophagus lid of Pacal (K'inich Janaab' Pacal in Mayan;
lived 603-683 CE) is one of the staples of ancient astronaut belief,
and it provides a good example of many of the failings of the theo-

ry. First, it is not considered in context. Ancient astronaut believers ignore the rest of the sarcophagus, and pay no attention to any other piece of Maya art and what such art might tell us about the image. This lack of context and selective choosing of evidence is endemic in pseudosciences, which look at isolated pieces of data and assemble broad theories based on them. In addition pseudoscientists seldom look at scholarly studies of the evidence they use to support their claims. In the case of Pacal, the textual evidence from the Temple of the Inscriptions, which scholars *can* read, gives a detailed account of his life and family. The fact that it never once mentions rockets, astronauts or doughnuts is ignored.

Further, we see the use of loaded language by pseudoscientists in describing the evidence. Note how here Silly implies that his interpretation is obvious "if only we take the time to look" and claims that the interpretation of the sarcophagus lid as representing a man in a rocket "has been verified by numerous world-famous scholars." Presumably he is referring to fellow pseudoscientists (Erich von Däniken, perhaps?),[83] since mainstream scholars of the Maya, world famous or not, reject the ancient astronaut explanation and instead argue, based on Maya iconography from a wide variety of sources, that the image reflects Pacal between life and death, which is an appropriate subject for a tomb lid.[84]

The rejection of coincidence is another common pseudoscientific technique, here presented in the fact that the death mask of Pacal is made of blue jade, the same general color often used to represent gods in the Hindu pantheon. Blue is found widely in the natural world, and Silly gives no reason why it should be considered significant here. Of course, since he tends to see every round shape in the world as representing an extraterrestrial doughnut (including the round shapes on the sarcophagus lid), perhaps we should not be surprised. He does at least give a reason for Pacal to be riding a spaceship, and so in this way has moved beyond the typical ancient astronaut believers, who seem to have little interest in what the image means, only what they think it shows.

The implication or accusation of scholarly conspiracies is common in pseudoscience. In his description of the phenomenon, Martin Gardner describes pseudoscientists as having two primary characteristics: they work alone and have a tendency towards paranoia.[85] Gardner's work on pseudoscience must be considered in its historical context, however; one commentator has noted that it "provides a flavour of the immense optimism surrounding science in the 1950s."[86] With this in mind, we must reassess Gardner's first claim, since in the 1950s the ability to disseminate information was far more restricted than it is today, and this played a significant role in isolating not merely pseudoscientists but anyone with interests outside of the mainstream. Today, with the internet and various forms of electronic publishing available, it is possible to form large social groups quickly and easily, and one of the results of this is that various pseudoscientific subcultures have emerged. What does remain generally true of Gardner's remarks is that pseudoscientists are still isolated outside the scientific mainstream, which evaluates claims based on evidence, not popularity.

The tendency towards paranoia among pseudoscientists is not uncommon, and in some sense it is understandable, since pseudoscientific theories do provoke strong emotions from scholars. The response to Velikovsky's theories has been seen by many scientists themselves as unduly harsh, in part because the early 1950s were a period when science was under attack from several quarters.[87] But the hostility between scientists and pseudoscientists is hardly one-way. Because pseudoscientists see the scientific mainstream as a rival for possession of a singular and exclusive truth, they often begin their presentations with attacks on the establishment. Consider the introduction to *Chariots of the Gods?*:

> It took courage to write this book, and it will take courage to read it. Because its theories and proofs do not fit into the mosaic of traditional archaeology, constructed so laboriously and firmly cemented down, scholars will call it nonsense and

put it on the Index of those books which are better left un-
mentioned.[88]

This statement comes *before* his ideas are presented, mind you, so
von Däniken was clearly anticipating a fight with scholars before his
book was even read. This may have been because memories of the
Velikovsky controversies were still fresh at the time, but we should
also note that those were far from the first arguments to take place
between pseudoscientists and mainstream scholars. In the 19th and
20th centuries, a large number of pseudoscientific ideas were proposed linking Native Americans with any of a number of cultures of
the Old World, as well as to sunken continents such as Atlantis and
Mu/Lemuria. Just as modern creationists and ancient astronaut believers do today, the adherents of these movements presented their
ideas to scholars and were soundly rebuffed. One of the more famous examples was our old friend Augustus le Plongeon and his
wife Alice. Unfortunately, as we have seen, le Plongeon also claimed
to be able to read the Maya hieroglyphs, and when he was challenged on his readings, responded with frustration and hostility:

> But who are these *pretended authorities?* Certainly not the doc-
> tors and professors at the head of the universities and colleg-
> es in the United States; for not only do they know absolutely
> nothing of ancient American civilization, but judging from
> letters in my possession, the majority of them refuse to learn
> anything concerning it…. The so-called learned men of our
> days are the first to oppose new ideas and the bearers of the-
> se. This opposition will continue to exist until the arrogance
> of self-conceit of superficial learning that still hover within
> the halls of colleges and universities have completely van-
> ished….[89]

More recently, Graham Hancock has remarked:

> Human history has become too much a matter of dogma,
> taught by professionals in Ivory Towers, as though it's all
> fact. Actually, much of human history is up for grabs. The

further back you go, the more that the history that's taught in the schools and universities begins to look like some kind of fairy story.[90]

Perhaps the most amusing and long-lasting caricature of the isolated, "Ivory Tower" scholar was created by Harold S. Gladwin, an amateur archaeologist who in 1947 created the character of Dr. Phuddy Duddy, a closed minded, arrogant representative of archaeological orthodoxy who was of course completely opposed to any form of independent thought.[91] The character was clearly inspired by the derogatory term "fuddy-duddy."

Lest we seem too harsh, though, it is worth asking: do scholars ever get trapped in orthodoxies of their own? The answer, sadly, is yes. The resistance to Schliemann that I mentioned earlier is a good example, and represents an almost pseudoscientific fixation on a paradigm. And in this way pseudoscience does provide a perverse sort of service, if only to show us what to beware of. The fact is that the reality of scientific personalities sometimes contrasts with the ideal of science. Kuhn notes that in cases of scientific revolutions, there is often a group of scholars who hold to the old paradigm despite the evidence supporting the new, and despite the anomalies that the old paradigm cannot explain.[92] And scholars also sometimes proceed deductively when they should be looking at a problem inductively. A good example of this from Assyriology is the Temple State hypothesis, in which scholars of the Early Dynastic Period in Mesopotamia constructed a hypothesis that claimed temples controlled most land in Sumerian society, translating texts from the site of Lagash to support this theory rather than accepting what the texts actually said.[93] But let us not forget that in actual science and scholarship, even when particular researchers become too attached to paradigms, their fields should not.

Obviously, some pseudoscientists (and scientists) are more combative than others. A statement regarding Mark Lehner from Graham Hancock and Robert Bauval on the *Guardian's* Egypt website takes a refreshingly civil tone, remarking that "we now welcome a cordial and amicable atmosphere that will allow us all to work, separately and perhaps even together, toward the truth behind the Giza plateau, no matter what this truth may eventually turn out to be."[94] Whatever the quality of their ideas and logic, both Zecheria Sitchin and Immanuel Velikovsky do present them in a fairly reasoned tone, which may be why they have not enjoyed Erich von Däniken's commercial success (though they have done quite well). Since the quality of writing in *Chariots of the Gods?* is really quite poor, I suspect that the reason for its success is partly the implied conflict with authority that permeates the text. As it does in fiction, conflict in pseudoscience sells.

This does bring us to another interesting feature of pseudoscientists: Even as some insult scholars and complain about scholarly persecution, many also express a strong desire that their views be accepted by the scholarly community. They seem to want membership in what they regard as an exclusive club that won't let them in, though it's worth noting the case of Mark Lehner, whose early interest in the pyramids of Egypt was based on the mystical statements of Edgar Cayce, but who is now one of the leading mainstream scholars on the subject.[95] As Kuhn notes, scientific disciplines are also communities. It is the nature of communities to have some set of shared values that distinguishes them, and central to the scientific community is the acceptance of a shared set of training and experiences.[96] The paradigm of a field may be challenged through a scientific revolution, but only if the new paradigm is successful at explaining both anomalies *and* previous data does it then become the new focus of the community. This means that to join a scientific community, one must either accept the old paradigm or replace it with a new one that actually works better. Because they believe in different paradigms which have not passed the scrutiny of scholarship, it is hard for pseudoscientists to join scientific communities and be taken seriously. Lehner only became a member of the community of Egyptologists when he stopped using Cayce's visions as evidence, studied Egyptology in detail at Yale, and began work within the paradigms used by Egyptologists, which he has since helped to refine.

So the contradictory need for both martyrdom and validation makes the pseudoscientist an odd figure, but we must remember that not all the aims of pseudosciences are the same; each movement has different goals. Creationists, for example, tend to seek political power, hoping to control the curriculum of schools in an effort to restore the authority of religious institutions over secular life; they are a feature of the modern sociological phenomenon of religious fundamentalism. Ancient alien believers seem to be most interested in support from popular culture, some possessing as well an almost messianic belief that their extraterrestrials will return (or are already here) and vindicate them. Other pseudosciences each have their own particular goals as well. What ancient doughnut theory is up to probably relates to snacking.

4.9

Here we see again the common problem in pseudoscience of reading meaning into things that may not have meaning, and of drawing conclusions from these readings. The classic Freudian joke about this is that sometimes a train in a tunnel really is just a train in a tunnel. We also see again a tone of writing frequently found in pseudoscientific works that is absent from mainstream scientific literature, even that which is produced for the general public. Pseudoscience is concerned not merely with evaluating evidence and providing empirical explanations for it, as science and scholarship are, but also with giving a sense of meaning to the evidence and the world. This means that pseudoscientific texts often read like adventure stories, and Silly's excitable tone here is quite similar to the that found in works like Stan Hall's *Tayos Gold: The Archives of Atlantis*.[97]

5.1

Here we see again the writing style of the travelogue, common in pseudoscientific books; see Graham Hancock's work for another example.[98] The "famous philosopher" being referred to by Silly is actually the comedian Steve Martin, who in the 1970s made jokes about his difficulties in France because the French were unwilling to accommodate his inability to speak their language. Since ethnocentrism is a common feature of pseudoscientists, it's not surprising to see it here, although to my knowledge Silly is the first to speculate that France was not created by the French. His wonder at the beauty of the Louvre is understandable, since it is an extraordinary building, but a little historical research would have told him that it was constructed over several centuries by French kings, a fact documented by both written sources and archaeology.

5.2

Silly's description of the law code of Hammurabi makes careful use of several facts about the stela in order to guide the reader to a particular interpretation, which he then claims is the only possible one. The stela is made of black diorite, and it is true that there is little native stone in Mesopotamia, but we also have ample evidence for widespread trade throughout Mesopotamian history and we know that stone was imported for monuments just such as this one. As to carving the hard stone, while it is true that this would be difficult with bronze tools, it would not be so difficult with equally hard stone.

The script is not some alien writing system but a deliberate effort at archaizing in order to make the laws seem older than they were. The earliest cuneiform writing was not read left to right, but top to bottom, and far from being "completely different," many of the signs on Hammurabi's stela are easily recognizable to even a first-year student of Akkadian, and all have later equivalents in the cuneiform system, something that Silly could easily have learned by consulting even the most basic of books on the subject. This illustrates an important point about many pseudoscientists: they regard education as an impediment to understanding, not a means toward it, and so fail to do even the most elementary research into the material they put forward in support of their ideas, relying instead on first impressions and hunches. Silly makes this clear here, but we can see a similar anti-educational slant from authors like von Däniken and Hancock, who accuse scholars of bias because their interpretations differ from the pseudoscientists. Von Däniken tells us:

> Excavations, old texts, cave drawings, legends, and so forth were used to construct a working hypothesis. From all this material an impressive and interesting mosaic was made, but it was a product of a preconceived pattern of thought into which the parts could always be fitted, though often with cement that was all too visible. An event must have happened in such and such a way. In that way and no other. And lo and behold—if that's what the scholars really want—it did happen that way.[99]

To say that this is an inaccurate description of how archaeology and history are actually done is to be too kind, but to say that it has a tone hostile to education and research is quite accurate. A similar hostility to scholars and the way they work has been expressed by Hancock, who stated the following in an interview for BBC Horizon:

> [Archaeologists] deprive [a site] of all mystery and render it as boring and predictable as possible, I think it would be nice if orthodox scholars approached it [Tiwanaku in Bolivia] with a slightly more generous and a more open attitude, and at least a willingness to be amazed, rather than writing that off at the outset.[100]

Precisely why archaeology and history are "boring and predictable" is a mystery to those of us who have spent our lives studying them, but as Fagan notes, the purpose of archaeology is to make things clearer, and if that makes them predictable, that is rather the point. On the other hand, relying on hunches and mystery often leads you to see what you want to see, as Silly demonstrates in abundance; he's hardly alone.

5.3

I am not aware of any other case where an author cites a quote from the same quote that they are citing, much less the same paragraph, so Silly may be an innovator here, though not really in a good way. It does bring to mind the case of Ayn Rand, who apparently had the interesting quirk of quoting her fictional hero John Galt as an authority on philosophical points, rather than herself, as though the mention of his name rather than her own carried additional gravitas,[101] and the practice of pseudoepigripha, common in the ancient world, of assigning your own ideas to a famous person from the past (pseudo-Paul and Paul in the question over authorship of the Pauline letters in the New Testament, for example). While it is not uncommon to cite your own previous work in scholarship, it does run the risk of making your arguments circular, though not sweet like a doughnut.

The glass pyramid that stands in front of the Louvre is hardly ancient, having been completed as part of a new entrance to the museum in 1989. But since establishing and using a consistent dating system for their evidence is not a hallmark of pseudoscientists, it's probably too much to expect Silly to do so here.

5.4

Again we see Silly practicing the common pseudoscientific technique of ascribing to his critics things they have not actually said; note that he does not actually cite anyone here. Von Däniken makes a similar claim, stating that the civilization of Egypt appeared suddenly and without precedent.[102] In fact, archaeologists have uncovered large amounts of information about both early Mesoamerica and early Egypt. In the case of Egypt, this shows continuous cultural evolution from as far back as 6000 BCE for nomadic herders and 5500 BCE for agriculture, not to mention Paleolithic activity for several hundred thousand years before that.[103] The evolution of pyramid construction is also well-documented, with a clear pattern that moves from tombs called mastabas to step pyramids to true pyramids (including one pyramid that failed and another that was corrected halfway through its construction).[104] The idea that ancient civilizations appeared from nowhere is common in archaeological pseudoscience, in part because pseudoscientists frequently do not do basic background research before making their claims, and in part because it suits the narrative they want to make, that only by miraculous events did these civilizations come into being.

Silly also makes an argument for radical diffusionism, which was a major point of the pseudosciences of the 19th and much of the 20th century, particularly with regard to the peopling of North and South America. Diffusionism argued against the idea found in the mainstream scholarship of the day that "civilization" had arisen independently in different parts of the world, and stated that there must have been contact between the Egyptians and the Maya, for example, since both peoples built pyramids. Diffusionism is based on the idea that it is only possible for human beings to invent things once, though diffusionists never explain why this is the case. But again, apart from their basic shape, the pyramids in Egypt have little in common with those in Mesoamerica. Egyptian pyramids evolved to have smooth sides and were never meant to be climbed, while those in Mesoamerica are temple platforms with stairs to the top. Further, the Egyptian pyramids are far, far older than those in Mesoamerica. The only overlap in dates (though again, not in shape) are between early Mesoamerican examples and the pyramids in the Sudan.

The use of linguistics to support diffusionism also has a long history, though its methodology is questionable. Linguists connect the members of language families by looking not only at similar words, but also similar structures. Thus, Semitic languages share the use of three-consonant roots for most verbs and their related nouns, for example. The fact is that any two languages will have some similar words or sounds, even in cases where the languages clearly cannot be related. The Norwegian indefinite plural suffix –ene is identical to the Sumerian animate plural suffix –ene, but this does not indicate a direct relationship between the two. For an example of pseudoscientific linguistics, see Barry Fell's *America B.C.: Ancient Settlers in the New World.*[105]

Pyramid at Chichen-Itza, Mexico (Photo by the Author)

Pyramid at Palenque, Mexico (Photo by the Author)

Pyramid at Giza, Egypt (Photo by the Author)

Pyramids at Jebel Barkal, Sudan (Photo by the Author)

More recent scholarship has taken a more nuanced view of the whole diffusionist-parallel evolution debate. We can no more argue that all cultures existed in complete isolation until the modern era than we can that they all derived from a single source (unless you want to go back to the original proto-humans of Africa). Sometimes there is evidence of limited contact, as with the prevalence of the sweet potato, native to South America, throughout Polynesia prior to contact with the Europeans. Since potatoes, unlike coconuts, cannot float across the ocean without dissolving, this would seem to indicate that at least one Polynesian ship reached the New World at some point, something the Polynesians were perfectly capable of doing. But if this contact occurred, it had little other long-term impact.[106]

5.5

The dating of the Sphinx is based on inscriptional evidence as well as its context in the broader arrangement of monuments on the Giza plateau, which date to the 4th Dynasty of Egyptian history (c. 2575-2465 BCE). Khafre is presented as the most likely candidate because the Sphinx is in association with his funerary monument, but there is a later tradition that tells of his predecessor Khufu (the builder of the Great Pyramid) doing repairs on the Sphinx, which might indicate either it came from earlier in the dynasty or that it was carved over a longer period of time and that Khafre merely took advantage of its proximity to his own monument.[107] Given that the Khufu inscription is much later (c. 600 BCE), it should be viewed with some caution.

The argument about the dating of the Sphinx has created a loosely affiliated group whose main argument is that it must be far earlier than c. 2500 BCE, the traditional date accepted by Egyptologists. John Anthony West, a believer in catastrophism who also has argued that the famous "face" on Mars is evidence of an ancient spacefaring civilization, brought up what appear to be peculiar weathering patterns on the body of the Sphinx, which, following another catastrophist named Schwaller de Lubicz, he attributed to a massive flood. Since no such flood is attested in Egyptian records (the Nile generally being quite regular in its seasonal rise and fall), de Lubicz and West after him argued that it must have taken place before Egyptian civilization as we know it emerged in the 4th millennium BCE. At this point the geologist Robert Schoch entered the picture, and in 1991 he argued that the weathering was caused by rainfall, which in Egypt we know was far heavier in prehistoric times than historic ones. Thus, Schoch concluded, the Sphinx must have been carved about 7000 BCE.

This is evidence, and it must be taken seriously, but what followed is a classic example of pseudoscience. West and Schoch have been joined by Robert Bauval and Graham Hancock, who proposed, based on astronomical alignments, that Schoch's date is too late, and that the Sphinx was actually carved in c. 10,500 BCE (remember their whole catastrophe argument). Now, the trouble with astronomical alignments is that they can be read any way you want them to, to give any answer you want. This is why they are so commonly used by pseudoscientists, since they are perfect for the defense of an overly rigid paradigm—all you have to do is mention math and astronomy and it looks like your "data" is beyond question. But since this sort of "evidence" is really just modern speculation, it is useless unless corroborated by other ancient material, which neither Bauval nor Hancock has been able to produce.

But you can see why they would be so taken by Schoch's argument for an early date of the Sphinx, as Silly has done here, especially since geology is a "hard" science and therefore considered by some to be more trustworthy than the social sciences and humanities. Unfortunately, Schoch's view is not shared by all geologists, and two others who have extensive experience in Egypt, K. Lal Gauri and James Harrell, have publically disputed his claims, arguing that the weathering patterns can be explained by the normal conditions on the Giza plateau since 2500 BCE. What this means is that the geological evidence for an early Sphinx is uncertain at best.

So what do we do? Here is where scholars and pseudoscientists go their different ways. Scholars argue that given the preponderance of evidence of 4th dynasty building activity on the Giza plateau, and the complete lack of evidence for any sufficiently complex civilization in Egypt in either 7000 BCE or 10,500 BCE, logic dictates that we agree with Gauri and Harrell. The pseudoscientists, on the other hand, simply ignore the evidence of the geologists who disagree with them, and ignore the lack of other archaeological evidence that the Egyptologists have noted. This is clearly symptomatic of placing one's paradigm over the evidence, one of the classic features of pseudoscience. Silly clearly falls into this camp.[108]

The lesson of the Sphinx dating controversy, therefore, is this: it is a perfect example of the failure of pseudoscience to evolve, and its inability to undergo a Kuhnian scientific revolution. Geological evidence that the Sphinx is thousands of years older than the 4th dynasty constitutes an anomaly, and following Kuhn, it is such anomalies that can provoke a crisis and eventually a scientific revolution, in this case a complete change in how we understand Egyptian (and with it ancient) history. But West, Schoch, Bauval and Hancock stop with the anomaly, failing completely to take the next step that any legitimate scholar would take, which is to see if the anomaly can be explained under the current paradigm. Since a reputable alternative to the early date has been reached by Gauri and Harrell, we need to ask if the weathering patterns actually constitute an anomaly at all.

The second failing of the pseudoscientists is actually more serious, but it is also characteristic of pseudoscience in general. Remember that in order for a paradigm shift to occur as a result of a scientific revolution, the new paradigm must do more than simply explain anomalies. It must also explain all of the old data that the old paradigm explained, and do so at least as well as the old paradigm. In the case of the dating of the Sphinx, the fact that the pseudoscientists make no real effort to do this dooms their entire enterprise. Because the current Egyptological paradigm does a far better job of accommodating all the old data than its proposed replacement, there is no reason to replace it, especially since Gauri and Harrell's explanation of the weathering patterns on the Sphinx explains the supposed "anomaly" as well. This is the way science works.

And V.R.Y. Silly? Well, he and the other pseudoscientists are up to something else entirely.

5.6

Here we see the presentation of "data" to support Silly's ideas. That the pyramids of Egypt took a great deal of effort to build is not in dispute, but the notion that they appeared suddenly is simply inaccurate. As I noted above, there was in fact a long period of development behind them. Like other pseudoscientists, Silly simply dismisses the technology of the Old Kingdom Egyptians as primitive, ignoring the fact that you can cut and move limestone blocks with stone tools and a number of related technologies that we have attested from Egypt.

Pseudoscientists have also frequently claimed that the pyramids, and the Great Pyramid in particular (which they frequently confuse with the slightly smaller pyramid of Khafre),[109] have either some mystical alignment with celestial objects or that they are the remnant of some long-forgotten technology that we cannot explain. To support these sorts of contentions, the dimensions of the Great Pyramid are often cited as somehow related to astronomical phenomena or objects. Silly's assertion is merely one of these; von Däniken repeats the old argument that "the area of the base of the pyramid divided by twice its height gives the celebrated figure π = 3.14159".[110] The problems are numerous here, however. The outer casing material of the Great Pyramid has long been gone, taken for other building projects, so we cannot actually say exactly what the original dimensions of the structure were. How, then, can we make such calculations to begin with?[111] The fact is that you can always play with numbers to get what you think are meaningful results; the same is true of astronomical alignments. For Tiahuanaco, Graham Hancock follows the ideas of Arthur Posansky to make his claim that the site should actually be dated to about 15,000 BCE.[112] But both Hancock and Silly ignore the *provenience* of the organic material that was used to produce the scholarly carbon-14 dates, which to an archaeologist is all-important.[113] There is also the problem that claiming that astronomical alignments are important to a site can only really work if you can also explain *why* those particular alignments were important to the builders, since any set of points on Earth can be aligned with something in the heavens. Other than supporting Hancock's rigid pseudoscientific paradigm, why are the Tiahuanaco alignments meaningful, and how does he demonstrate this? Like Silly, he doesn't, preferring to leave it a mystery.

This brings us to the question of evidence as it is used (or more accurately, misused) in pseudoscience. In any observational field, including history and archaeology, the quality of data is particularly important, since it is not possible to perform new experiments that test any individual data point. Thus, measurements, translations, etc., must be of the highest quality possible, performed by individuals with extensive training in the field; one should no more rely on an amateur here than one would for brain surgery. Unfortunately, because pseudoscience relies heavily on ambiguity and mystery, this sort of accuracy is often lacking in pseudoscientific studies. It should be noted that the *quantity* is not the same as the *quality* of data. Many pseudoscientists gather impressive volumes of information to support their claims. Those who deal with the ancient world often present material from cultures worldwide, for example. But despite this, a closer examination of their data reveals several major qualitative problems.

Pseudoscientists are usually very selective about what features of a piece of data they choose to use as evidence. For the ancient world, dates are used when convenient, if at all, making it possible to present two artifacts as related even if they are separated by several millennia (again, the pyramids of Egypt and Mesoamerica are a good example). Further, when comparing data, similarities may be emphasized and differences downplayed, but the criteria for this distinction are not given, even though when interpreting data, such criteria are essential.

There is also the question of simple accuracy. Pseudoscientific texts are often riddled with errors of fact that even the most basic fact-checking should correct before publication. Some authors, such as von Däniken, are worse about this than others, but the presence of such errors sends the message that pseudoscientists are sloppy and even deceitful.

Further, the data sets of pseudoscientists are rarely complete; by this I mean they rarely address all the types of data available in an academic field. So the pseudoscientists who argue for ancient astronauts, for example, produce data from certain types of artifacts (monumental architecture, artwork), and from certain types of texts (myths and legends), but not from others (we search in vain for the much larger quantity of evidence from excavations of the dwellings of commoners, or daily administrative texts, which exist in abundance). Now, it could be argued that the pseudoscientist is merely using the texts that are relevant to their theory, but remember that they are not working within the current paradigm for their field but rather are proposing an entirely new one. Since one of the purposes of a paradigm is that it allows the scholar to work without having to reinvent the field with every research project, it is not necessary to deal with all extant data in a study that is operating within a current paradigm. But when the researcher is proposing an entirely new paradigm (a scientific revolution), as pseudoscientists do, it is a requirement that they do test this new paradigm against *all* relevant data. Repeating the same arguments about the same data, such as the pyramids, does not a scientific revolution make.

Related to this is the problem of previous sources. When they consider previous studies at all, pseudoscientists often cite only a few, and generally only those which they believe support their assertions. This means that they will accept a single view of a problem as the only view, implying that the subject is settled when in fact it may not be. In *Fingerprints of the Gods*, Graham Hancock builds his views about early maps from Charles Hapgood's book *Maps of the Ancient Sea Kings* and Simon Bethon [sic] and Andrew Robinson's *The Shape of the World: The Mapping and Discovery of the Earth*, reducing his references for these early chapters to little more than a list of "ibid."[114] One result of this is that Hapgood's theory of "earth crust displacement" is presented as established fact, without discussion of any controversy that might surround it, even though many geologists have rejected the theory.[115]

The question of depth is related to this. Science is in a constant state of evolution between scientific revolutions. New data becomes available and new interpretations are proposed, and it is a requirement that scholars keep up with these in their field. They are also expected to familiarize themselves with as much data in their field as possible, even if it does not seem to directly relate to their current research (this is one reason it is important for professors to teach general survey courses to college freshmen). This allows both better work within the paradigm and modification of that paradigm as necessary, to say nothing of an increased ability to spot anomalies should these appear. But because of their emphasis on rigid paradigms and deductive reasoning, pseudoscientists tend to disregard or ignore data that does not seem relevant to their interests. Thus, as I noted above with Ezekiel and his supposed spaceship, they focus on single data points without concern for broader context.

The underlying reason for these failures is the issue I have raised several times: for the pseudoscientist, data exists to *support* their paradigm, not to *evaluate* it, and so evidence does not need to be carefully tested for accuracy, does not need to be complete, and does not need to be deep or considered in context. This results in claims that often misstate facts, but this too is unimportant to the pseudoscientist; defending the rigid paradigm against perceived attack becomes their all-consuming passion.

5.7

Here we see the common pseudoscientific technique of trying to control and win arguments rather than reach for objective truth through the examination and interpretation of evidence. By appropriating Sagan's speculations about the possibility of ancient extraterrestrial visitations (which Sagan noted were mere speculations, lacking positive evidence),[116] ancient astronaut believers have turned what began as an interesting question about astronomy into a cultural phenomenon where they have been permitted to set the parameters. In this way archaeologists and historians are put on the defensive. Instead of being required to present positive evidence for their assertions, the pseudoscientists simply connect negative evidence with proof, and then claim that it is scholars who must demonstrate that they are wrong rather than the pseudoscientists having to demonstrate that they are right. This allows the pseudoscientist to make statements like Silly's, or the following by Giorgio A. Tsoukalos:

> If it turns out that the ancient astronaut theory is wrong, I will be the first one to say: "Okay, no problem." But so far, certain pieces of evidence have not been debunked.[117]

Note the wording here: *"If* it turns out…" (emphasis mine). This implies that ancient astronaut belief has already been proven and that it is now the dominant paradigm of archaeology and history, and that to challenge it would require a new paradigm *without* ancient aliens in it. In fact, since it is the ancient alien believers who are proposing a new paradigm, the burden of proof falls on them, not on the archaeologist or the historian, and if the ancient alien believers are unable to present positive evidence that accounts for *everything* that the current paradigms account for, as well as the anomalies they claim exist (which usually turn out not to be anomalies after all), their attempt at a scientific revolution must be considered a failure. It is not the responsibility of the scientist to debunk the pseudoscientist, but rather the responsibility of the pseudoscientist to demonstrate why his or her interpretation is correct. If you want to call yourself scientific, you have to act like a scientist, not a lawyer.

5.8

Silly ends his presentation on a positive note, with wording that brings to mind a bit of narration from Ed Wood's infamously and delightfully bad film *Plan 9 from Outer Space*. Silly adds the insistence that he is right and that his critics are wrong, demonstrating a feature he shares with many other pseudoscientists; in fact his final line bears significant resemblance to a statement made by Erich von Däniken, who tells us:

> There is no doubt that my hypothesis will be accepted by the mainstream. It may be a question of just five or ten years. Sooner or later we will have contact with someone out there, and the thinking changes completely. And one day they will return, and if you don't listen to Erich von Däniken, you will have the *shock of the gods*.[118]

Note also the almost religious character of Silly's final arguments. Because pseudoscience is so concerned with telling a good story, and with giving a sense of meaning to its audience, such a tone is not surprising. In this way, pseudoscience really is more like religion than anything else, even as pseudoscientists claim to be purely scientific. Consider the following quotes from the History Channel's *Ancient Aliens* series:

> We have logic, we have intuition, we have our senses. And to limit ourselves through the scientific route is a mistake. We can send people on the moon, we can plan journeys to Mars, we can make cell phones, but we haven't explained the fundamental questions that we should be dealing with: Where do we come from? What are we supposed to do here? And where are we going? We really have not come with the answers and we know no better than the ancients that we're trying to explain.[119]

> Exploring the ancient astronaut theory is the ultimate quest to find out where did we come from, how did it all begin,

and most importantly, *who are we?* And, the ancient astronaut theory has the capability of answering all those questions. [120]

These are very good questions (I've asked them myself, and you may have too), but statements like "what are we supposed to do here?" and "who are we?" are metaphysical and religious, not concerned with facts but with meaning. Science cannot answer them. Rather, it can only give us a foundation so that we may use other means to try.

1 1970, 14-15.

2 1995, 437.

3 1950.

4 1976, 204-235.

5 While it is possible that scholars who disagree with the theories of von Däniken and his followers chose not to participate in the *Ancient Aliens* production, which after all had one of the leading proponents of ancient astronaut belief as its consulting producer, this does not mean that their views could not have been better included in what claimed to be an impartial production.

6 For a recent discussion of the subject of extraterrestrial life, see Monte Ross, *The Search for Extraterrestrials: Intercepting Alien Signals* (2009).

7 1970, 139-140.

8 1970, 140-141.

9 1970, 100.

10 See von Däniken 1970, 13.

11 See Kruger and Dunning 1999. A nice summary can also be found in the TED talk by J. Marshall Shepherd at https://www.ted.com/talks/j_marshall_shepherd_3_kinds_bias_that_shape_your_worldview.

12 The classic work on this subject is Thomas Kuhn's *The Structure of Scientific Revolutions* (1962), which I'll be citing a lot.

13 1995, 9-10.

14 See James and Thorpe 1999, 58-76 for a detailed discussion of the Poleshift theory.

15 See Jones 1967, 25-45 and Wood 1985, 87.

16 1979, 181.

17 Holocaust denial surely ranks as one of the most pernicious of the pseudosciences, and because it is frequently used to justify racism and anti-Semitism it requires particular attention. For a detailed discussion of Holocaust deniers and an analysis of how professional historians study the Holocaust and have concluded that it indeed did occur, see Shermer 1997, 2002, 175-241.

18 1997, 2002, 144.

19 See, for example, Sagan 1963; Shklovskii and Sagan 1966; Sagan 1973; Ross 2009.

20 1976, 1-4. The tongue-twister in this sentence is unintentional, I assure you.

21 1999. Diamond does a nice job of debunking older racist arguments regarding the rise and success of complex technological cultures. Intelligence and skin color, for example, have never been shown to be related.

[22] Again, see Diamond (1999, esp. table 5.1). In the Middle East, available domesticates form an impressive list: emmer and einkorn wheat, barley, pea, lentil, chickpea, flax, muskmelon, olive, sheep, and goats.

[23] It's always good to look at the original. I used the *Modern Library* edition. *The Origin of Species by Means of Natural Selection* was published in 1859.

[24] See Powell 1998, 176-179. It is worth noting that the asteroid impact theory (or Alvarez Hypothesis) is not universally accepted in paleontology. See in particular the work of Gerta Keller (http://gkeller.princeton.edu/publications). At the least, however, an asteroid impact would have resulted in a *very* bad day.

[25] von Däniken 1970; Clarke 1973.

[26] Webster 1999, 25.

[27] 1970, 16.

[28] 1976, 54.

[29] 1970, 22.

[30] James and Thorpe 1999, 192-194; Bloch 2012, 38-39.

[31] See, for example, Diamond 1999, 18-22.

[32] Alexander 2008, 37.

[33] The Smithsonian website has a nice telling of the story. See Damrosch, 2007.

[34] The question of left-brain versus right-brain function and perception is larger than we have room for here. In some cases of stroke or other brain trauma, however, one or the other side of the brain might be diminished or shut down, creating a radically different view of reality. Two interesting cases are that of Phineas P. Gage, who had a spike go through his brain after an industrial accident in 1848 and so lost his left frontal lobe but survived, and Jill Bolte Taylor, who temporarily lost the use of her left brain during a stroke in 1996. See https://www.ncbi.nlm.nih.gov/pmc/articles/PMC1114479/ and Bolte Taylor 2008.

[35] Blavatsky 2009, 9-11.

[36] See Said 1979.

[37] Churchward 1926, 2007, 17-21.

[38] Prabhavananda and Manchester 1957, 118.

[39] von Däniken 1970, 45-47.

[40] For a nice summary of the dialogues, see de Camp 1954, 1970, 3-10.

[41] 1954, 1970, 10-16.

[42] Donnelly 1882, 1976, 1-2.

[43] Really, the best work about Atlantis remains L. Sprague de Camp's *Lost Continents: The Atlantis Theme in History, Science and Literature* (1954, 1970). De Camp was a science fiction writer and so knew something about how

to spin a good tale, and his style is far more readable than the usual scholarly tome.

[44] Some basic surveys to get you started are *Sharer and Traxler* (2006) and *D'Altroy* (2015).

[45] 2009, 31.

[46] 1976, 168-170.

[47] 2009, 214.

[48] Thomsen 1984, 2001, 11. One scholar, I.M. Diakonoff, quips halfseriously that "there are as many Sumerian languages as there are Sumerologists" (1976, 99).

[49] Sitchin 1976, 170, figures 84 and 85.

[50] Labat 1999, 46-47 and 154-155; Borger 2004, 147 and 360.

[51] 2009, 214.

[52] 2002, 206.

[53] For the Minoans, see Driessen and MacDonald (1997) and MacGillivray (2001).

[54] See Hancock 1995 in particular, but also Bauval and Gilbert, 1994 and Flem-Ath and Flem-Ath 1995, 2012. For the rebuttal, see James and Thorpe 1999, 58-64 and 228-230.

[55] See Fritze 2009, 218.

[56] von Däniken 1974; Ferris 1974, 58.

[57] Story 1976, 19-20; Fritze 2009, 209-210. Some acknowledgement was given in von Däniken's second book, *Gods from Outer Space* (1971, x), though without stating what these other authors had specifically proposed. A look at *The Gold of the Gods* shows lots of claims and virtually no citations, making it really difficult to double-check von Däniken's assertions.

[58] Hall 2006.

[59] See esp. 2006, 21-30.

[60] Story 1976, 88-90.

[61] See Blavatsky 2009.

[62] Churchward 1926, 2007. The text of the Book of Mormon can be found at https://www.lds.org/scriptures/bofm?lang=eng.

[63] von Däniken 1970, 9-12, 29-30.

[64] I would be remiss if I didn't mention the evidence that we are becoming more civilized in the long run. Steven Pinker gives a nice summary of his views on this question in his TED talk "Is the world getting better or worse? A look at the numbers." (Pinker, 2018) Hopefully this debate will not end anytime soon, as a less violent world would be better for all of us and talking about violence honestly is a good start if we're going to stop it.

[65] Joseph F. Blumrich, Patent Number 3789947, registered 2-5-1974. See http://www.spaceshipsofezekiel.com/other/US_Patent_3789947-omnidirectional_wheel.pdf.

[66] 1970, 37-40.

[67] 2002, 236; though the Minoans have been associated with Atlantis by some.

[68] von Däniken 1970, 16-17.

[69] For the scholarly debates, see the discussion in James and Thorpe 1999, 322-336. Some scholars have argued for astronomical alignments, others that they lines are processional paths, and still others that they relate to watercourses.

[70] *Ancient Aliens: Chariots, Gods, and Beyond.* Bonus Feature from Ancient Aliens: The Series. Season 1, Episode 6; 2010; 1:23:33.

[71] See Hapgood 1966, 1996, 72-77; von Däniken 1970, 15; Hancock 1995, 4-5.

[72] McIntosh 2000, 67-68 and 94-96; Dutch 2012.

[73] See note 24 above.

[74] Aldred 1988; Burridge 1995.

[75] Hawass 2010, 54.

[76] See McIntosh 2000, 10.

[77] Feder 2004, 205-214.

[78] For the Maya, their intricate dating system, and their take on 2012, see David Stuart's *The Order of Days* (2011), whose final chapter talks at length about the real Maya and the imaginary Maya of pseudoscientists and New Age thinkers.

[79] For a review of the problems in American history textbooks, see James W. Loewen, *Lies My Teacher Told Me* (2007).

[80] Bendici 2009.

[81] 1970, 97.

[82] For le Plongeon, see Wauchope 1962, 7-20.

[83] See von Däniken 1970, 100-101.

[84] James and Thorpe 1999, 86; Feder 2002, 212.

[85] 1952, 8-15.

[86] Erickson 2005, 150.

[87] Storer 1977, 36-37; a thorough study of these events has been done by Gordin (2012).

[88] von Däniken 1970, vii.

[89] le Plongeon quoted in Wauchope 1962, 73; italics in cited text.

[90] *The Return.* Ancient Aliens: The Series. Season 1, Episode 5; 2010; 1:26:25.

[91] Wauchope 1962, 71-73.

[92] 1962, 149-151.

[93] Foster 1981, 237.

[94] Hancock and Bauval 1998.

[95] See Lehner's book *The Egyptian Heritage* (1974).

[96] Kuhn 1962, 167-168.

[97] 2006.

[98] This does make *Fingerprints of the Gods* (1995) a most entertaining read.

[99] 1970, 13.

[100] From Fagan 2012. Fagan states that the quote was taken from Hancock's own website, but Hancock's page was not operative when I tried to access the quote there. I leave it to the reader to decide on the citation's accuracy.

[101] Heller 2009, 336. See Rand 1964, 13.

[102] 1970, 74.

[103] See Wendorf and Schild 1976; Hoffman 1979.

[104] A nice, if slightly dated, summary of the Egyptian pyramids can be found in I.E.S. Edwards' book *The Pyramids of Egypt* (1947, 1961).

[105] 1976.

[106] Davies 1979, 200-201.

[107] See James and Thorpe 1999, 221.

[108] A nice summary of this whole debate can be found in James and Thorpe 1999:214-231, with citations. See also Hancock 1995:275-458.

[109] See the repeated examples in the History Channel's *Ancient Aliens* series. To be fair, scholars sometimes make the same mistake, which is less excusable. See, for example, the image in James and Thorpe (1999, 201), which is clearly the pyramid of Khafre photographed from the top of the actual Great Pyramid, not the Great Pyramid itself.

[110] 1970, 77.

[111] Story 1976, 75.

[112] For Hancock's position on this debate, see his website at https://grahamhancock.com/carbon-dating-hancock/.

[113] See Fagan 2012.

[114] Hancock 1995; Hapgood 1966, 1996; Bethon and Robinson 1990.

[115] Krause 1996; Fritze 2009, 217-218.

[116] See Shklovskii and Sagan 1966, 453-464 and Sagan 1973, 204-205.

[117] Giorgio A. Tsoukalos, *The Return*. Ancient Aliens: The Series. Season 1, Episode 5; 2010; 1:25:27.

[118] *Ancient Aliens: Chariots, Gods, and Beyond*. Bonus Feature from Ancient Aliens: The Series. Season 1, Episode 6; 2010; 1:27:22. I have put what seems to be verbal emphasis in italics.

[119] Robert Bauval, *The Return*. Ancient Aliens: The Series. Season 1, Episode 5; 2010; 1:26:48.

[120] Giorgio A. Tsoukalos, *The Return*. Ancient Aliens: The Series. Season 1, Episode 5; 2010; 1:27:22. I have put what seems to be verbal emphasis in italics.

Whither Pseudoscience?

The fault, dear Brutus, is not in our stars but in ourselves.
—William Shakespeare, *Julius Caesar*

Science cannot answer. This seems an odd statement to make in the 21ˢᵗ century, an age when science has explained so much, has made so many things possible, including more comfortable and longer lives for so many. It seems stranger still in a society that is dedicated to the factual truths that science is designed to understand. But perhaps this statement is more telling than we might at first think. Science, of course, is not a *thing* but a *way*, and we often forget this. It is a way to understand the physical world around us, a set of techniques, an intellectual construction designed to make sense of observed phenomena using logic and evidence. Correctly it is the scientific *method*, a verb, not a noun. Believing it is a thing gets us into trouble because all too often we confuse the results of that method, the data and paradigms and tentative conclusions it reaches, with the method itself, with how we came to those tentative conclusions.[1] The astronomical observation of a supernova by itself is not science, but the way we theorize about how the supernova came about is. This is true whether our explanation is the action of a god or the laws of physics observed through a telescope. What has changed between ancient Mesopotamia and now are the explanations of the phenomena we then produce, what we call paradigms. We now reject the explanation of Mesopotamian gods moving the heavens because modern physics does a better job explaining what we observe, but in their day, the scholars in Babylon were doing science.

Pseudoscience, on the other hand, is all about the *conclusion*, not the method, about interpreting data to support a particular idea or explanation and discarding or dismissing any evidence that does not say what the rigid paradigm wants it to say. The paradigm rigidity that is characteristic of pseudoscience therefore makes it a very different thing than science, and it is difficult to speak of a "pseudoscientific method" unless it is as a sales pitch, an effort to gain adherents. Further, the pseudoscientist does not merely choose data selectively, does not merely misinterpret it and misrepresent it to fit a rigid paradigm, but they also tend to create needlessly complex and ad hoc explanations as they build their narratives. For reasons never given an infallible God creates organisms filled with flaws and yet it is called "intelligent" design. After travelling many light years and leaving no concrete traces of their passing (not even a single piece of litter from a technology greater than our own), aliens or doughnuts, rather than Egyptians and Mesoamericans, build pyramids. Why do we need such explanations when simpler and more consistent ones are available? Why discard Occam's Razor so readily?

This is what strikes scholars as odd. It is not that pseudoscience isn't entertaining—quite the opposite. I thoroughly enjoy fictional stories about ancient aliens, from the seminal *2001: A Space Odyssey* to the sadly short-lived television show *Defying Gravity*. Again, I recall a conversation with my grandfather Franklin Roach when he described his participation in the Air Force study of UFOs, and his excitement when he thought he might have actually found one (it turned out to be a satellite disintegrating on reentry).[2] To put it bluntly, the rejection of ancient alien belief by the scholarly community isn't because we are hostile to the idea, but rather because we are doing what we have been trained to do, to be skeptical and careful. Again, as Marcello Truzzi and Carl Sagan put it, "Extraordinary claims require extraordinary evidence."

We live in an age of wonders, and learning about them is exciting. My cell phone has an app that lets me zoom out from the Earth to the furthest reaches of the known universe, and which identifies planets around other stars, an ability almost unthinkable only a few decades ago. So why is pseudoscience still so popular? It cannot be that science or other scholarship is boring.

Two thoughts come to mind, both of which reflect the nature of our brains. Cognitive scientists have begun to speculate why it is so easy to fool the human mind. Daily we are treated to stories of people misled in everything from business to politics, so why should science prove an exception? To try and answer this question, let us return to our old friend Darwin and his amazing insight. Put simply, Darwin argued that traits which favor a particular organism in a particular environment will be more likely to be passed on to later generations of that organism's descendents. But what is most important for our purposes is that Darwin said nothing about whether those adaptive traits included an objective view of reality. In fact, it is easy to see conditions where fiction would prove advantageous over objectivity. If an organism develops an outsized fear response against predators, for example, it will respond and overreact to many false alarms, and it will run from many predators that are not there. But this also means that it will always run from the threat that *is* there, and in so doing improve the odds that it and its offspring will survive and propagate. When it comes to interpreting much of the data from your senses, false positives can literally save your life, while false negatives can get you killed.

So how *do* we understand what we see? The psychologist Jerome Bruner distinguishes two modes of interpretation, two ways we look at the world around us, which he calls the "logico-scientific" and the "narrative." The logico-scientific deals with "universal truth conditions," and the narrative with "particular connections between two events."[3] I propose that Bruner's two modes also reflect the distinction between *logos* and *mythos,* between truths of fact and truths of meaning, and that ultimately these two modes allow us to deal with differing degrees of complexity.

The logico-scientific mode, and the scientific method, are well suited to problems that repeat themselves, that can be tested, and that can be described with logic and with mathematics. These tend to be simpler problems that lend themselves to natural laws such as those of physics and chemistry. We understand them better by breaking them down into their simpler components. But narratives are more versatile, allowing us to consider far more complex, unique, and untestable problems.[4] And while they are used in the presentation of science, narratives are not a part of the scientific method for one reason in particular: they require authors. An experiment or an observation is simply that. It tells what it tells; it is merely evidence and does not explain itself. When a scientist makes an observation, she must interpret it, must try to make sense of it. The rules of science tell her to bias toward the simplest explanation that accounts for all the data, and to follow the data where it leads. Only when this inductive process is done can a theory be created and a narrative built to present it. Deduction follows as the theory is tested.

The presence of an author, however, elevates causality to the status of intention. The author chooses what is to be included and what is to be left out, what matters and what doesn't. And so when we assign an author to logico-scientific problems, we render science invalid, for if God or aliens or doughnuts are the cause of natural phenomena, then those phenomena become capricious, unpredictable, and even the laws of physics can change without warning. It is from this belief that physical miracles become possible, and because pseudoscience, like religion, ultimately rejects the logico-scientific in favor of the narrative, rejects natural causes in favor of an author, it cannot therefore be science.

A number of recent scholars have considered this question from the perspective of the biology and evolution of the human brain.[5] Put simply, they note that one of the most important survival advantages human beings have is our ability to cooperate in social groups, since individually we are vulnerable to any number of larger, faster, and better-clawed predators. Jonathan Haidt argues that a major feature of the human adaptive kit, and a major reason for our success as a species, is our ability to belong to groups. A result of this is that we are frequently willing to subsume our own interests, and our own beliefs, to those of the group.[6] This includes placing the opinions of the perceived leaders of these groups over what we observe, or even what we previously believed.[7] So if our neighbor, or simply any important authority figure, tells us that aliens built the pyramids to accommodate giant intergalactic doughnuts, then many will believe it even if that belief, as in this case, is patently absurd and contrary to what they actually see.

Further, we will defend these beliefs so long as we are surrounded by others who do likewise, and it is worth noting that successful pseudoscientists do not tend to operate alone. Note the emphasis on joining and belonging found in the website of an ancient alien organization:

> But **most importantly you will support the cause to spread the Ancient Astronaut theory by being a part of this pioneering organization.** With your participation, the cause will gain momentum to enlighten the world about mankind's cosmic past. Not only do we have the answers to your questions, but you will also **support the cause!** (boldface in original)[8]

In the marketplace of ideas, this is actually an advantage for the pseudoscientist, because science is ultimately not interested in adherence or followers or even finding answers that make you feel good. Rather, science seeks to understand the empirical world as it *actually is,* and honest scientists will accept even conclusions that make them uncomfortable. Remember that the scientific method is designed to explain *how* things work, not to explain *why*. The Big Bang theory argues only that the universe we live in came into existence some billions of years ago in a single explosion that created not only matter and energy but also time and space. It does not say *why* this occurred.

So pseudoscience resides in a narrative place, but with a flaw. It attempts to serve the function of religion, of myth, to bring us meaning and address the question of *why,* but it cloaks itself in the language of science to do this because, as Johannes Sloek argues, modern Western society has a positive view of scientific sounding texts, whether they are actually scientific or not.[9] We "moderns" regard truth as factual only, and we have lost our myths and so seek them in science, where they do not belong. Pseudoscientists tell exciting tales, wrapped in mystery. They promise the same sorts of answers once provided by our myths, the same meaning. God created the universe, and you and me, and because God is a sentient mind, there must be a purpose to our creation, a reason. Only through the intentional action of a divine figure can the physical world operate, and in this view, faith in a religious orthodoxy trumps what you observe.[10] Or aliens built pyramids and created us through genetic engineering, again, for a reason, giving us what the religious studies scholar Andreas Grünschloß calls "a *secular and ufological parallel to creationism.*"[11] In both cases we are part of a larger story, not merely the result of random chance and impersonal physical and chemical laws.

But is the appeal of pseudoscience (as opposed to its fraudulent claims) always a bad thing? It's easy to forget that a purely scientific worldview, while perhaps more factually accurate, also makes significant demands. It tells us that there is no safety net, that we cannot assume the presence of a greater intelligence giving our lives meaning. And so it is not unreasonable that many believe in the mystical, the religious, and that we seek things that look like sciences that can also reject the scientific, and in so doing allow us to keep meaning, purpose, and safety. We understandably want to have our cake and eat it too. In relation to believers in the pseudosciences of creationism and intelligent design, Philip Kitcher has discussed this with considerable sympathy:

> They know that the case launched against their cherished beliefs is clever, but they are also tempted by the thought that the cleverness is flawed. If others, recognizably more sympathetic to their faith, can point however vaguely to potential faults,

they will be grateful—and they will be disinclined to inspect too closely the gifts they are offered.[12]

And,

> To resist Darwin, or the enlightenment case that looms behind him, is hardly unreasonable if what you would be left with is a drab, painful, and impoverished life.[13]

In other words, we should not be so quick to assume that believers in Atlantis or ancient aliens or creationism are simply suffering from an inability to think logically, but rather should consider that they believe what they do because it helps them deal with the vicissitudes of their lives. This includes the importance of belonging to a supportive social group. In some ways this may be a fear response (creationism in particular is governed by the fear that modern, scientific society requires a rejection of God), or it may simply be a desire for an exciting and mysterious narrative, the kinds of tales our ancestors once told around the campfire. Remember that it's easy to be comfortable and tenured in the Ivory Tower and talk about a purely mechanistic world in the abstract, but harder still, particularly in the increasingly unfair and unequal modern world, to do so when the deck is stacked against you. When science seems so sterile, is it any wonder that people find comfort in pseudoscience, which after all promises the certainty of science together with the certainty of faith?

There are also those who might argue that pseudoscientific beliefs have more in common with religious ones. But there is a crucial difference here. Religions, at their core, perform functions not found in either science or pseudoscience. First, religions are less about *logos* than they are about *mythos*. Put another way, they are about *why*, where science is about *how*. Second, while it is central to the scientific method to try and move from more complex to simpler explanations of things (Occam's Razor again: the simplest explanation is more likely correct), religion embraces complexity and is designed for it. This is why religion confronts the kinds of complex social and moral problems that science struggles to explain, and it is why religion remains essential to human life even as science expands what it can tell us. It is why many scientists are deeply religious, with no degradation in the quality of their work.

So in order to identify pseudoscience, we have to consider whether the explanation is addressing *how* or *why*. Pseudoscience addresses *how*, just as science does—How were the pyramids built? By who? How was the Earth created? By who or what? How do vaccines or antibiotics work? But when we ask *why*, we are entering the realms of religion and philosophy and the humanities, of things whose physical causes and effects might be identifiable scientifically, but whose meanings are increasingly complex, where Occam's Razor is not only not useful but can actually be a hindrance. When asking *why* questions we must be prepared to embrace complexity, not reduce it, must embrace the fact that each event or circumstance is unique and no single explanation can possibly cover them all. If the scientific *how* tells us the way our car engine works, the philosophical and religious *why* looks for the reasons we took the drive. Both are important, but they are not the same thing. Pseudoscience is "crackpot" not only because it fails to effectively address the *how*, but also because it fails to really consider the difference between the *how* and the *why*. It may closely resemble religion, but by claiming that it reveals scientific truths, it winds up failing at both.

But never forget that the problem is not confined to pseudoscience. Scientists can make this *how* and *why* mistake also, can wind up claiming or denying meaning instead of explanations, demanding that meaningful things be proven as facts if they are to have value. On the other side of things, religious believers and philosophers can lose the deeper value of their narratives by demanding that those narratives be accepted as literal facts, again a mistake of substituting *how* for *why*, as happens in creationism. To remember this distinction between *how* and *why* allows us the best of both science and religion, of both simplicity and complexity, while to confuse the two benefits no one.

Finally, we must also remember that there are going to be those things which do not fit neatly into any of these categories. In such cases we have phenomena that are so complex that scientific investigation is impossible (though aspects of them can be looked at by scientific means) but which also defy easy religious explanation (though again, they may contain religious symbolism); they do not fit any single theology or approach. These include experiences which occur in unique and singular ways and which are nearly impossible to put into words or art by those to have them. They cannot be repeated experimentally. As an example, I note the phenomena of near-death and other profound spiritual experiences. Those who have had such experiences (including the author of this book) find them nearly impossible to describe or explain, even as there is no question but that they happen and that they are frequently life-changing. And while some may argue that studies of these sorts of things are pseudoscience, it is noteworthy that they generally do not challenge any scientific paradigm and lack a common paradigm of their own. Others may argue that they are a sort of pseudoreligion, but they also lack any central theology. Like the experiences themselves, they are generally outside of language, and sometimes we just have to admit that there are certain things we do not fully understand.

I close with a note of caution, repeating the observation of Donald Fritze: pseudoscience *can* be dangerous.[14] It's hard to imagine ancient astronaut or Atlantis believers ever posing any real threat, and by and large the followers of these ideas are perfectly nice people. They do provide some entertainment, and as I noted in the introduction, this is one of the major reasons I have written *Doughnuts of the Gods*. But let us not be fooled: there is a dark side to pseudoscience. It frequently encourages academic fraud, and it is hostile to education, both because a well-educated audience can all the more easily debunk these theories and because those who are well-educated are less likely to obey an orthodoxy without examining it first (though there are, sadly, exceptions to this rule; we are also, remember, very influenced by the beliefs of those around us).

Other dangers are more serious. Because one of the features of pseudoscience is its reliance on the orthodoxy of a seductive and exciting idea, there is the very real risk that such ideas will move from the realm of opinion and into the realm of action. Some pseudosciences make apocalyptic claims, and apocalyptic thinking always carries with it the temptation to do violence, whether to yourself or others. The tragedy of the Heaven's Gate cult, where 39 people committed suicide in 1997 because they thought that the approaching Hale-Bopp comet was accompanied by an alien space ship coming to pick them up, is an example of this, as are the suicides of people who were convinced that the world would end in 2012 because the Maya said so; in fact, the Maya said no such thing.[15] So let me be clear about this: ancient aliens and doughnuts are funny; suicide is *not*.

Further examples of dangerous pseudoscience include efforts to use political power to enforce the orthodoxy of a pseudoscientific belief. Perhaps the most chilling example is Nazi racial ideology, which was based largely on pseudoscientific ideas and carried to its horrific and logical conclusion in the Holocaust.[16] Modern Holocaust deniers, Neo-Nazis and others who espouse racial hatred spread their ideas quickly on the internet.[17] In another case, the Soviet biologist Trofim Lysenko managed to convince Josef Stalin to ban the use of Darwinian evolutionary theory in favor of his own Lysenkoism in 1948. The enforcement of Lysenko's pseudoscientific ideas by the Soviet state led to several biologists being executed or sent to gulags, and the study of biology in the Soviet Union was crippled for decades afterward. This likely contributed to many deaths from famine.[18] In more recent times, the use of deceptive "studies" to impact research into the dangers of smoking almost certainly contributed to many early deaths, and the rise of "fake news," particularly on the internet, has been used to promote particular political candidates and agendas, endangering the fundamental principle of an informed electorate that is essential to democracy.[19] As marijuana becomes increasingly legalized, we should all ask ourselves if similar misinformation about its safety, risks and benefits could affect this debate (whether pro- or con-), and whether we may someday pay a price for this if we aren't careful and honest about the science.[20]

Today we are also confronted by the anti-vaccination movement, which uses pseudoscience against one of the most effective disease-fighting tools ever created. As Caitlin O'Connor and James Owen Weatherall have noted, this can lead to both a public health crisis and needless deaths.[21] The use of political power to enforce their ideas is appealing to many pseudoscientists, and the efforts of creationists to use the courts to push their religious beliefs in public schools in the United States is troubling, to say the least.[22] All this should serve as a warning to scholars and the public to take the phenomenon of pseudoscience seriously, even in silly parodies like the one you are holding. Always remember that the most dangerous idea in the world is the one you *want* to believe, since this is the one that you are least likely to evaluate critically.

Pseudoscience is a fascinating thing, an odd mimic of science, unable to make significant and legitimate discoveries on its own but very good at pretending that it does. Contrary to some views,[23] it shares some features with science, following a similar process of paradigm formation. In the ways it is different, however (rigidity of paradigm and the influence of *mythos,* for example), pseudoscience results in a completely different method of understanding the empirical world and shares much with fundamentalist religion. Factually, it is regularly debunked, both at the level of methodology and at the level of evidence, but it will never go away. The reason for this is that the pseudosciences do not appeal to the rational parts of our brains but rather to the irrational, to feelings more than ideas, to the maintenance of a mystery rather than its resolution. Whether we are talking about Atlantis or ancient astronauts or intergalactic doughnuts, it is this that gives these ideas their power, and it is this that we must understand when we communicate with their adherents. Scholars can and do present accurate information about the world in ways that are meaningful (witness Carl Sagan's hugely popular series *Cosmos* or the television series *The Day the Universe Changed* by James Burke), but it is sadly also true that science and scholarship can come across as sterile as we seek to minimize bias and practice reasonable scholarly caution. Nonetheless, I can attest from my own teaching that the excitement for ancient history that has been a big part of my life can be transmitted to others, and if there is a lesson we may learn from pseudoscience it is this one: Never abandon the careful attention to facts and evidence and interpretation, but feel free to express the meaning, the *mythos*, that these things inspire in you. You don't need aliens or doughnuts or Hollywood misinterpretations of reality to make your world interesting. You just need yourself, and you need to be willing to share the fascinating real world with others so they will be less likely to get drawn into fictions that are presented as facts. Direct confrontation with pseudoscientists seldom works, and data and logic have their limitations when you are dealing with true believers, but there is no need to let pseudoscientists monopolize the excitement of learning. Ultimately a better understanding of yourself and a willingness to think critically are the best ways to approach these rather peculiar ideas about our world, and when necessary to protect yourself from them.

May the many flaws and silliness of ancient doughnut theory help to light your way.

---

[1] A nice and easy to follow presentation about how scientists work and deal with the mistakes they will inevitably make as they seek and refine their answers, and how making mistakes is an important part of the process, not a weakness, can be seen in this short TED talk by Phil Plait: "The secret to scientific discoveries? Making mistakes." (https://www.ted.com/talks/phil_plait_the_secret_to_scientific_discoveries_making_mistakes).

[2] For an account of this, see Roach 1999, 168-173.

[3] 1986, 11-12.

[4] This raises the interesting question of left-brain and right-brain specialization. Generally speaking, the left brain is good at detail work but does not tend to see the big picture, while the right brain is good at the big picture and directs the left brain to solve the details of problems. For a fascinating discussion of this, see the *Hidden Brain* episode with the psychiatrist Iain McGilchrist, "One Head, Two Brains: How the Brain's Hemispheres Shape the World We See." (https://www.npr.org/2019/02/01/690656459/one-head-two-brains-how-the-brains-hemispheres-shape-the-world-we-see).

[5] The fallibility of human reason has been known and studied for some time. A good look can be found in Gilovich 1991. More recent examinations of the problem include Gorman and Gorman 2016, Mercier and Sperber 2017, and Soloman and Fernbach 2017.

[6] 2012, 191. See also O'Connor and Weatherall 2019.

[7] Cohen 2003, 811.

[8] Legendary Times Website, 2019.

[9] 1996, 42.

[10] See Shermer 1997, 127-172.

[11] 2006, 18. Italics in original.

[12] 2007, 156.

[13] 2007, 160.

[14] 2009, 10.

[15] For the suicides, see Fritze 2009, 167-168 and Fraknoi et al 2013, 7. For the beliefs of the ancient Maya, see Stuart 2011, 310.

[16] For the relationship of Nazism to Atlantis belief, see Fritze 2009, 61.

[17] See the discussions in Shermer 1997, 2002, and Fritze 2009 for good summaries, with further references.

[18] Gardener 1952, 144-146; Gordin 2012, 81.

[19] See the discussion in O'Connor and Weatherall, 2019. On a personal level, I mention my friends Walter Burt, Bill Zorn, Perry Clinton, and Marty Martinez, all of whom were sickened and often died too young after smoking for many years. I doubt I am alone in this sort of loss.

[20] The marijuana debate is extensive and deserves a book of its own, perhaps several. I do note that smoking *anything*, however, from tobacco to marijuana to dried banana leaves, is putting stuff into your lungs that they were not designed to have there, and that newer strains of marijuana have higher levels of THC than the old ones (see https://archives.drugabuse.gov/rise-in-marijuanas-thc-levels). From a policy perspective, it seems to me that these risks should be considered against the massive costs of the so-called "War on Drugs," which has enriched drug lords and other criminal elements as it brings high levels of violence (McNamara, 2011). As an alternative to prohibition, we have the example of Portugal, which decriminalized drugs in 2001 and has seen a significant drop in drug-related problems since then.
(see https://www.theguardian.com/news/2017/dec/05/portugals-radical-drugs-policy-is-working-why-hasnt-the-world-copied-it
and http://www.emcdda.europa.eu/countries/drug-reports/2018/portugal_en).

[21] See the discussion in O'Connor and Weatherall 2019, 140-44.
The medical literature in favor of vaccines is huge, but here are a few starting points from the Centers for Disease Control: *What Would Happen If We Stopped Vaccinations?* https://www.cdc.gov/vaccines/vac-gen/whatifstop.htm and
Vaccines Do Not Cause Autism.
https://www.cdc.gov/vaccinesafety/concerns/autism.html

[22] A summary of this can be found in Shermer 1997, 154-172. For the history of creationism, a good starting point is Ronald Numbers' *The Creationists: From Scientific Creationism to Intelligent Design, Expanded Edition* (2006).

[23] See Bunge 2009.

Bibliography

Alcock, Leslie. 1971, 1973. *Arthur's Britain*. New York: Penguin Books.

Aldred, C. 1988. *Akhenaten, King of Egypt*. London: Thames and Hudson, Ltd.

Alexander, Caroline. "Stonehenge." *National Geographic*, June 2008, 35-59.

Ancient Aliens: Season One. 2010. The History Channel. DVD.

Armstrong, Karen. 2000. *The Battle for God: A History of Fundamentalism*. New York: Ballantine Books.

Bauval, Robert, and Adrian Gilbert. 1994. *The Orion Mystery: Are the Pyramids a Map of Heaven?* New York: three Rivers Press.

Beck, Martha. 2005. *Leaving the Saints: How I Lost the Mormons and Found My Faith*. New York: Crown Publishers.

Bendici, Ray. 2009. "Damned Interview: Dr. Kenneth L. Feder." Accessed 10-26-2012. http://www.damnedct.com/damned-interview-dr-kenneth-l-feder/

Berthon, Simon and Andrew Robinson. 1990. *The Shape of the World: The Mapping and Discovery of the Earth*. New York: Rand McNally.

Blavatsky, H.P. 2009. *The Secret Doctrine*. Abridged and annotated by Michael Gomes. New York: Penguin Group.

Bloch, Hannah. "The Riddle of Easter Island." *National Geographic*, July 2012, 30-49.

Blumrich, Joseph F. Patent Number 3789947, registered 2-5-1974. http://www.spaceshipsofezekiel.com/other/US_Patent_3789947-omnidirectional_wheel.pdf

Bolte Taylor, Jill. 2008. *My Stroke of Insight: A Brain Scientist's Personal Journey*. Viking.

Book of Mormon. Accessed 1-26-2019. https://www.lds.org/scriptures/bofm?lang=eng

Borger, Rykle. 2004. *Mesopotamisches Zeichenlexikon*. Alter Orient und Altes Testament 305. Münster: Ugarit-Verlag.

Brown, Peter Lancaster. 1976. *Megaliths, Myths and Men: An Introduction to Astro-Archaeology*. New York: Harper Colophon Books.

Bruner, Jerome. 1986. *Actual Minds, Possible Worlds*. Cambridge, MA: Harvard University Press.

Bunge, Mario. 2009. "The Philosophy Behind Pseudoscience." In *Science Under Siege: Defending Science, Exposing Pseudoscience*. Edited by Kendrick Frazier, 235-251. Amherst, New York: Prometheus Books.

Burridge, A. 1995. "Did Akhenaten Suffer From Marfan's Syndrome?" *Akhenaten Temple Project Newsletter* No. 3.

Centers for Disease Control and Prevention. *What Would Happen If We Stopped Vaccinations?* Accessed 2-23-2019. https://www.cdc.gov/vaccines/vac-gen/whatifstop.htm

_____. *Vaccines Do Not Cause Autism*. Accessed 2-23-2019. https://www.cdc.gov/vaccinesafety/concerns/autism.html

Churchward, James. 1926, 2007. *The Lost Continent of Mu*. Kempton, Illinois: Adventures Unlimited Press.

Clarke Arthur C. 1973. *Profiles of the Future: An Inquiry into the Limits of the Possible*. Popular Library.

Cohen, Geoffrey L. 2003. "Party Over Policy: The Dominating Impact of Group Influence on Political Beliefs." *Journal of Personality and Social Psychology* 85: 808-822.

D'Altroy, Terence. 2015. *The Incas* (2nd ed.). West Sussex: Wiley Blackwell.

Damrosch, David. 2007. "Epic Hero. How a self-taught British genius rediscovered the Mesopotamian saga of Gilgamesh after 2,500 years." Accessed 1-26-2019.
https://www.smithsonianmag.com/history/epic-hero-153362976/

von Däniken, Erich. 1970. *Chariots of the Gods?* New York: Bantam.

_____. 1971. *Gods from Outer Space.* New York: Bantam.

_____. 1974. *Gold of the Gods.* New York: Bantam.

Darwin, Charles. 1859 and 1871. *The Origin of Species & The Descent of Man.* New York: Modern Library.

Davies, Nigel. 1979. *Voyagers to the New World.* New York: William Morrow and Company.

de Camp, L. Sprague. 1954, 1970. *Lost Continents: The Atlantis Theme in History, Science and Literature.* New York: Dover Publications.

Diakonoff, Igor M. 1976. "Ancient Writing and Ancient Written Language: Pitfalls and Peculiarities in the Study of Sumerian." 99-121. *Assyriological Studies of the University of Chicago* 20. Chicago: University of Chicago Press.

Diamond, Jared. 1999. *Guns, Germs and Steel: The Fates of Human Societies.* New York: W.W. Norton and Company.

Donnelly, Ignatius. 1882, 1976. *Atlantis: The Antediluvian World.* New York: Dover Publications.

Driessen, Jan and Colin MacDonald. 1997. *The Troubled Island.* Aegaeum 17 Annales d'archéologie egéenne de l'Université de Liège et UT-PASP. Liège : Université de Liège, Histoire de l'art et archéologie de la Grèce antique. Austin: University of Texas at Austin, Program in Aegean Scripts and Prehistory.

Dutch, Steven. 2012. "The Piri Reis Map." Accessed 9-23-2012. http://www.uwgb.edu/dutchs/PSEUDOSC/PiriRies.HTM

Edwards, I.E.S. 1947, 1961. *The Pyramids of Egypt.* New York: Penguin Books.

Erickson, Mark. 2005. *Science, Culture and Society: Understanding Science in the 21st Century.* Cambridge: Polity Press.

European Monitoring Center for Drugs and Drug Addiction. "Portugal Country Drug Report 2018." Accessed 2-16-2019. http://www.emcdda.europa.eu/countries/drug-reports/2018/portugal_en

Fagan, Brian. 2012. "An Answer to Graham Hancock" Accessed 10-4-2012. http://www.hallofmaat.com/modules.php?name=Articles&file=article&sid=18

Feder, Kenneth L. 2002. *Frauds, Myths and Mysteries: Science and Pseudoscience in Archaeology.* 4th Edition. Boston: McGraw-Hill.

Fell, Barry. 1976. *America B.C.: Ancient Settlers in the New World.* New York: Pocket Books.

Ferreira, Susana. "Portugal's radical drugs policy is working. Why hasn't the world copied it?" The Guardian. Accessed 2-16-2019. https://www.theguardian.com/news/2017/dec/05/portugals-radical-drugs-policy-is-working-why-hasnt-the-world-copied-it

Ferris, Timothy. "Playboy Interview: Erich von Däniken." *Playboy*, Number 8, 1974, 51 ff.

Flem-Ath, Rand and Rose Flem-Ath. 1995, 2012. *Atlantis Beneath the Ice.* Revised and Expanded Edition. Rochester, Vermont: Bear & Company.

Foster, Benjamin. 1981. "A New Look at the Sumerian Temple State." *Journal of the Economic and Social History of the Orient* 24: 225-241.

Frair, Wayne. 2000. "Baraminology—Classification of Created Organisms." *Creation Research Society Quarterly Journal* 37, No. 2: 82-91. Accessed 5-25-2019. http://web.archive.org/web/20030618153040/http://www.creatio nresearch.org/crsq/articles/37/37_2/baraminology.htm

Fraknoi, Andrew, Kristine Larsen, Bryan Mendez, David Morrison, and Mark Van Stone. 2013. "Doomsday 2012 and Cosmophobia: Challenges and Opportunities for Science Communication." *Communicating Science: A National Conference on Science Education and Public Outreach ASP Conference Series.* Vol. 473: 3-11. J. Barnes, C. Shupla, J. G. Manning, and M. G. Gibbs, eds.

Fritze, Ronald H. 2009. *Invented Knowledge: False History, Fake Science and Pseudo-religions.* London: Reaktion Books.

Gardner, Martin. 1952, 1957. *Fads and Fallacies in the Name of Science.* New York: Dover Publications.

Gilovich, Thomas. 1991. *How We Know What Isn't So: The Fallibility of Human Reason in Everyday Life.* New York: The Free Press.

Goldsmith, Donald, ed. 1977. *Scientists Confront Velikovsky: Evidence Against Velikovsky's Theory of Worlds in Collision.* New York: W.W. Norton.

Gordin, Michael D. 2012. *The Pseudoscience Wars: Immanuel Velikovsky and the Birth of the Modern Fringe.* Chicago: University of Chicago Press.

Gorman, Sara E. and Jack M. Gorman. 2016. *Denying to the Grave: Why We Ignore the Facts That Will Save Us.* Oxford: Oxford University Press.

Grünschloß, Andreas. 2006. "»Ancient Astronaut« Narrations: A Popular Discourse on our Religious Past." *Marburg Journal of Religion* 11, No. 1, 1-25. Accessed 1-18-2014. http://archiv.ub.uni-marburg.de/mjr/art_gruenschloss_2006.html

Haidt, Jonathan. 2012. *The Righteous Mind: Why Good People are Divided by Politics and Religion.* New York: Pantheon.

Hall, Stan. 2006. *Tayos Gold: The Archives of Atlantis.* Kempton, Illinois: Adventures Unlimited Press.

Hancock, Graham. 1995. *Fingerprints of the Gods.* New York: Crown Publishers.

_____. "Position Statement on Carbon-Dating" Accessed 1-26-2019. https://grahamhancock.com/carbon-dating-hancock/

Hancock, Graham and Robert G. Bauval. 1998. "Statement Regarding Dr. Mark Lehner from Graham Hancock and Robert G. Bauval." Accessed 9-19-2012. http://www.guardians.net/egypt/statement7-98.htm

Hapgood, Charles. 1966, 1996. *Maps of the Ancient Sea Kings.* Kempton, Illinois: Adventures Unlimited Press.

Hawass, Zahi. "King Tut's Family Secrets." *National Geographic,* September 2010, 34-59.

Heller, Anne C. 2009. *Ayn Rand and the World She Made.* New York: Anchor Books.

Hoffman, Michael A. 1979. *Egypt Before the Pharaohs.* New York: Alfred A. Knopf.

James, Peter and Nick Thorpe. 1999. *Ancient Mysteries*. New York: Ballantine Books.

Jones, Tom B. 1967. *Paths to the Ancient Past: the Applications of the Historical Method to Ancient History*. New York: The Free Press.

Keller, Gerta. "Gerta Keller Publications." Accessed 1-26-2019. http://gkeller.princeton.edu/publications

Kitcher, Philip. 2009. *Living With Darwin: Evolution, Design, and the Future of Faith*. Oxford: Oxford University Press.

Krause, Steve. 1996. "Hapgood's Theory of Earth Crust Displacement." Accessed 9-18-2012. http://www.skrause.org/writing/papers/hapgood_and_ecd.shtml

Kruger, Justin, and David Dunning (1999). "Unskilled and Unaware of It: How Difficulties in Recognizing One's Own Incompetence Lead to Inflated Self-Assessments." Journal of Personality and Social Psychology. 77 (6): 1121–1134.

Kuhn, Thomas S. 1962. *The Structure of Scientific Revolutions*. Chicago: University of Chicago Press.

Labat, René. 1999. *Manuel d'Épigraphie Akkadienne*. 6th Edition. Paris: Librarie Orientlaliste Paul Geuthner.

Larsen, Kristine. 2008. "This I ~~Believe~~ Understand: The Importance of Banning the B-Word from Science." Astronomy Education Review Issue 2, Volume 6:118-126. Accessed 3-8-2019. http://access.portico.org/Portico/#!journalAUSimpleView/tab=P DF?cs=ISSN_15391515?ct=E-Journal%20Content?auId=ark:/27927/pgg3ztf9wzn

"The A.A.S. R.A." Legendary Times Website. Accessed 1-16-2019. http://www.legendarytimes.com/index.php?op=page&pid=29

Lehner, Mark. 1974. *The Egyptian Heritage*. Edgar Cayce Foundation.

Loewen, James W. 1995, 2007. *Lies My Teacher Told Me.* New York: Simon & Schuster.

MacGillivray, J Alexander. 2001. *Minotaur: Sir Arthur Evans and the Archaeology of the Minoan Myth.* London: Pimlico.

Mack, John E. 1994. *Abduction: Human Encounters with Aliens.* New York: Charles Scribner's Sons.

McIntosh, Gregory. 2000. *The Piri Reis Map of 1513.* Athens, Georgia: The University of Georgia Press.

McNamara, Joseph D. 2011. "The Hidden Costs of America's War on Drugs." The Journal of Private Enterprise 26(2), 97-115.

Mercier, Hugo and Dan Sperber. 2017. *The Enigma of Reason.* Cambridge: Harvard University Press.

Mufson, Beckett. 2018. "Apparently, Some People Believe the Earth Is Shaped Like a Donut." *Vice.* Accessed 4-6-2019. https://www.vice.com/en_us/article/mbyak8/apparently-some-people-believe-the-earth-is-shaped-like-a-donut-1

National Institute on Drug Abuse. "A Rise in Marijuana's THC Levels." Accessed 2-16-2019. https://archives.drugabuse.gov/rise-in-marijuanas-thc-levels

Numbers, Ronald. 2006. *The Creationists: From Scientific Creationism to Intelligent Design, Expanded Edition.* Cambridge: Harvard University Press.

O'Connor, Cailin and James Owen Weatherall. 2019. *The Misinformation Age: How False Beliefs Spread.* New Haven: Yale University Press.

O'Driscoll, Kieran and John Paul Leach. 1998. "'No longer Gage': an iron bar through the head: Early observations of personality change after injury to the prefrontal cortex." BMJ. 317(7174): 1673–1674.

Pinker, Steven. 2018. "Is the world getting better or worse? A look at the numbers." Accessed 3-12-2019. https://www.ted.com/talks/steven_pinker_is_the_world_getting_better_or_worse_a_look_at_the_numbers?language=en

Plait, Phil. 2019. "The secret to scientific discoveries? Making mistakes." Accessed 3-23-2019. (https://www.ted.com/talks/phil_plait_the_secret_to_scientific_discoveries_making_mistakes)

Plato. 1952. *The Dialogues of Plato.* J. Harward, translator. Chicago and London: Encyclopedia Britannica, Inc.

Powell, James Lawrence. 1998. *Night Comes to the Cretaceous: Dinosaur Extinction and the Transformation of Modern Geology.* New York: W.H. Freeman and Company.

Prabhavananda, Swami, and Frederick Manchester, translators. 1957. *The Upasnishads.* New York: New American Library.

Rand, Ayn. 1964. *The Virtue of Selfishness.* New York: Signet.

Roach, Franklin Evans. 1999. *Musings and Memoirs of Franklin Evans Roach.* Santa Fe, NM: Private publication.

Ross, Monte. 2009. *The Search for Extraterrestrials: Intercepting Alien Signals.* Berlin: Praxis Publishing, Ltd.

Sagan, Carl. 1963. "Direct Contact among Galactic Civilizations by Relativistic Interstellar Spaceflight." *Planetary Space Science* 11: 485-498.

_____. 1973. *The Cosmic Connection: An Extraterrestrial Perspective.* New York: Anchor Press, Garden City.

_____. 1977. "An Analysis of *Worlds in Collision.*" In *Scientists Confront Velikovsky: Evidence Against Velikovsky's Theory of Worlds in Collision.* Goldsmith, Donald, ed. 41-104. New York: W.W. Norton.

Said, Edward. 1979. *Orientalism.* New York: Vintage Books.

Sharer, Robert J. and Loa P. Traxler. 2006. *The Ancient Maya* (6th, fully revised ed.). Stanford, California: Stanford University Press.

Shepherd, J. Marshall. 2018. "3 kinds of bias that shape your worldview." Accessed 3-12-2019. https://www.ted.com/talks/j_marshall_shepherd_3_kinds_bias_th at_shape_your_worldview

Shermer, Michael. 1997, 2002. *Why People Believe Weird Things.* Revised Edition. New York: St. Martin's Griffin.

Shklovskii, I. S. and Carl Sagan. 1966. *Intelligent Life in the Universe.* San Francisco: Holden-Day.

Sitchin, Zecharia. 1976. *The 12th Planet.* New York: Avon Books.

Sloek, Johannes. 1996. *Devotional Language.* Trans. H. Mossin. Berlin and New York: de Gruyter.

Sloman, Steven and Philip Fernbach. 2017. *The Knowledge Illusion: Why We Never Think Alone.* New York: Riverhead Books.

Storer, Norman W. 1977. "The Sociological Context of the Velikovsky Controversy." In *Scientists Confront Velikovsky: Evidence Against Velikovsky's Theory of Worlds in Collision.* Goldsmith, Donald, ed. 29-39. New York: W.W. Norton.

Story, Ronald. 1976. *The Space-Gods Revealed. A Close Look at the Theories of Erich von Däniken.* London: New English Library.

Stuart, David. 2011. *The Order of Days: The Maya World and the Truth About 2012.* New York: Harmony Books.

Thomsen, Marie-Louise. 1984, 2001. *The Sumerian Language: An Introduction to its History and Grammatical Structure.* Mesopotamia 10. Copenhagen: Akademisk Forlag.

Vedantam, Shankar, Rhaina Cohen, Tara Boyle, and Jennifer Schmidt. 2019. "One Head, Two Brains: How the Brain's Hemispheres Shape the World We See." *Hidden Brain.* Accessed 2-24-2019. https://www.npr.org/2019/02/01/690656459/one-head-two-brains-how-the-brains-hemispheres-shape-the-world-we-see

Velikovsky, Immanuel. 1950. *Worlds in Collision.* New York: Pocket Books.

Ward, Peter and Donald Brownlee. 2003. *Rare Earth: Why Complex Life is Uncommon in the Universe.* New York: Copernicus Books.

Wauchope, Robert. 1962. *Lost Tribes and Sunken Continents: Myth and Method in the Study of American Indians.* Chicago: The University of Chicago Press.

Webster, Donovan. "Journey to the Heart of the Sahara." *National Geographic,* March 1999, 2-33.

Wendorf, Fred and Romauld Schild. 1976. *Prehistory of the Nile Valley.* New York: Academic Press.

Wood, Michael. 1985. *In Search of the Trojan War.* New York: Facts on File Publications.

Zapozooz, Yakinzoz. 2156. *What the Hell am I doing on This Planet?* Alpha Centauri: Alien Pastry Publications.

Index

www.ingramcontent.com/pod-product-compliance
Lightning Source LLC
Chambersburg PA
CBHW072228190626
46809CB00017B/1524